Growth and Decay of Coral Reefs

Growth and Decay of Coral Reefs: Fifty Years of Learning describes how coral reefs have alternately flourished and declined over the last 50 years and the dynamics of these changes. The study is based on recordings at 30 different locations along the Sudanese coast, visited by the author between 1971 and 1973.

Beyond the Red Sea's desert shores lie some of the richest and most diverse coral reefs on our planet. Over a thousand species of reef fishes, matched by a similar abundance of living corals, creating habitats scientists were only just beginning to understand. The complexity of the inter-relations was truly mesmerizing. A single intervention, such as removal of a key species, could cause the whole community to collapse. Healthy corals were transformed into green weed-smothered reefs, accompanied by the loss of both corals and fish.

Based on the author's observations of how knowledge and perspectives have changed over the last 50 years, this book highlights lessons learned from historical records that may help maintain and reestablish coral reefs in the years to come.

Topics covered include:

CORAL REEF FISH
Fish surveys

CORAL REEF PROFILES
Coral growth rates
Coral distribution
Corals on 'Cousteau garage'

CORAL THREATS
Climate change
Coral bleaching
Coral diseases
Coral sponges
Terpios hoshinota
Coral predators
Coral urchins

CORAL – ALGAE

CORAL RESEARCH
and more . . .

Peter J. Vine is a marine biologist and author with special interests in coral reef ecology, marine conservation, photography, and film-making. An adjunct lecturer at the National University of Ireland, Galway, he established one of the country's first commercial fish farms – a move that led to activities in the USA, Europe, and the Middle East. Returning to Ireland, a series of authorship commissions led to the establishment of three media companies involved in book publishing, film production, and project development in support of a series of award-winning national pavilions at World Expos in Spain, China, South Korea, and Italy. Peter is married to Paula and has three daughters, Catriona, Sinead, and Megan, all of whom, with their respective partners, live and work in Ireland.

Red Sea, typical of its time, included four new species and a new genus. Ichthyologists were having a field day, describing hundreds of fish that were new to science.

From 1974 to 1976, I dived on 30 different reefs in the central Red Sea and marvelled at what greeted me underwater. The survey included morphology and ecology of (1) fringing reefs at

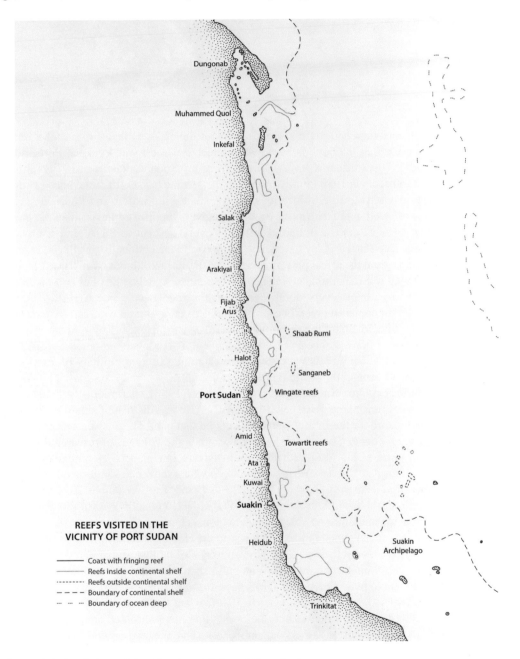

Figure 1.1 Reef sites visited in the vicinity of Port Sudan.

© Vine

Chapter 1

Introduction

This is a book about change and more specifically about the changes that impact coral reefs. Much of the data on which the book is based was collected while I was living in Port Sudan in the early to mid-1970s. I began writing descriptions of my Sudanese dives in 1970 and first published the reef transects in 1980 (Vine & Vine 1980). As time passes, the relevance of some of these observations, at over 30 reef locations, grows. Clear blue waters and flourishing reef communities were enjoyed, more than half a century ago, with youthful enthusiasm, excitement, and a stimulating sense of discovery that greeted our every descent to these almost pristine underwater habitats. Even in those days, we knew that we were exceptionally privileged to have front-row seats at this great theatrical performance of the natural world: our own "Blue Planet."

Could it last? Would it last? Would, as some journalists suggested, hoards of crown-of-thorns starfish (CoTS) destroy almost every coral reef across the Indo-Pacific ocean? Little did we know that the exuberant applause greeting each act in this riveting drama would, over the coming half-century, be overwhelmed by an ominous pall of death, triggered by climate change. Despite apparent pockets of resistance and evidence for seasonal recovery among a few species, there was no denying that coral reefs were steadily succumbing to a multiplicity of attacks on tropical marine habitats around the world (Hass 1971).

The eminent marine explorer, Cmd Jacques-Yves Cousteau (1953), was among the first to alert the world to the devastation taking place beneath the waves. He was distressed to find reefs dying before his eyes and feared for the future. Could it really be that bad? The answer, unfortunately, was "yes." Was there no hope? The answer lies somewhere in the volumes of research that have been published by some of the most dedicated biologists on our planet. Encouragingly, faint glimmers of light illuminate the darkness, evoking hope where evidence suggests otherwise. A dwindling number of scientists at Glasgow's COP26 in October to November 2021 remained cautiously optimistic in the face of a growing view that we have already missed this particular boat. They argue that we might just be able to adapt to global change before we are overwhelmed by it. But will nature also have the resilience to adapt to the changing conditions that threaten to shut down the show? We need it to do so – for our own world to stabilise and offer a viable future.

Before we go further, I must state that this is an unfinished story and one that surprises even the most pessimistic biologists. There actually IS hope for our reefs, given time and care. This story looks back to those heady days of the early 1970s, when Commander Jacques Cousteau made his dire predictions, and it delves into the events that have unfolded since then. Importantly, it reminds us of the progress that is being made to survive the crisis – working with nature rather than against it. It begins in the relatively early days of SCUBA diving, when almost anything that one collected in such remote coral zones stood a chance of being "new" to science. Thus, my paper (Vine 1972a) on the tube worm members of the Spirorbinae in the Sudanese

DOI: 10.1201/9781003335795-1

Acknowledgements

I am pleased to thank Prof. Yousif Abu-Gideiri at the Department of Zoology, University of Khartoum, for his encouragement and assistance in carrying out fieldwork in the region. The original studies were carried out while I was employed as director of Suakin Marine Laboratory (University of Khartoum). I am grateful to Dr. Douglas Allan and Paula Vine for assistance with diving. I also acknowledge Dr. J.E. Randall, Dr. Rupert Ormond, Dr. Dirar Nasr, Prof. Hans Mergner, Prof. Johannes Schroeder, Dr. B. Smith-Vaniz, Dr. Gerry Allen, Dr. Peter Rowe, Mr. R. Moore, and Dr. Stephen Head, with whom I discussed the fieldwork. I am also pleased to acknowledge the assistance of Ms. Fiona Martin for her work on illustrating this publication.

Finally, the book owes its genesis to Dr Najeeb Rasul whose encouragement was both critical and practical and without which this book would not have been written.

Contents

Dr. John Randall (1924–2020), or simply "Jack," published over 900 papers and described 799 new fish species. He wrote 13 regional fish guides and maintained active research right into his 90s. Indeed, his last joint publication, "Endemic Fishes of the Red Sea" (Bogorodsky & Randall 2019), was an important contribution to the Red Sea fauna. I first met Jack at the Bernice P. Bishop Museum in Hawaii, where he curated a large collection of mainly coral reef fish. Later, he came to stay as our guest in Port Sudan. I dived with him on many occasions and was always impressed by his aquatic skills as both an observer and hunter of Red Sea reef fishes. Never without a Hawaiian sling, he fired his narrow spear to enter and leave his target fish on the same side, leaving the opposite side unblemished and suitable for his renowned specimen photography. Jack inspired at least as many young biologists as he described new species. He regarded personal communications as a key ingredient in his impressive scientific output. In my own case, he was a valued mentor and generous contributor of his unique knowledge on Red Sea fish. I am indebted to him for helping to create a list of fishes in Suakin Harbour that appears in this book. He was one of the most inspirational people I have ever known, and it is an honour to dedicate this volume to his memory.

Thank you, Jack; rest in peace.

Designed cover image: © Hans Sjöholm
Science illustrations by Fiona Martin

First published 2024
by CRC Press/Balkema
4 Park Square, Milton Park, Abingdon, Oxon, OX14 4RN

and by CRC Press/Balkema
2385 NW Executive Center Drive, Suite 320, Boca Raton FL 33431

CRC Press/Balkema is an imprint of the Taylor & Francis Group, an informa business

© 2024 Peter J. Vine

British Library Cataloguing-in-Publication Data
A catalogue record for this book is available from the British Library

Library of Congress Cataloging-in-Publication Data
Names: Vine, Peter, author.
Title: Growth and decay of coral reefs : fifty years of learning / Peter J. Vine.
Description: Boca Raton, FL : CRC Press, 2023. | Includes bibliographical
    references and index.
Identifiers: LCCN 2023010999 (print) | LCCN 2023011000 (ebook) |
    ISBN 9781032371955 (hardback) | ISBN 9781032371962 (paperback) |
    ISBN 9781003335795 (ebook)
Subjects: LCSH: Coral reef ecology—Red Sea. | Corals—Red Sea.
Classification: LCC QH541.5.C7 V56 2023 (print) | LCC QH541.5.C7 (ebook) |
    DDC 577.7/89—dc23/eng/20230606
LC record available at https://lccn.loc.gov/2023010999
LC ebook record available at https://lccn.loc.gov/2023011000

ISBN: 978-1-032-37195-5 (hbk)
ISBN: 978-1-032-37196-2 (pbk)
ISBN: 978-1-003-33579-5 (ebk)

DOI: 10.1201/9781003335795

Typeset in Gill Sans
by Apex CoVantage, LLC

Growth and Decay of Coral Reefs

Fifty Years of Learning

Peter J. Vine

CRC Press
Taylor & Francis Group
Boca Raton   London   New York

CRC Press is an imprint of the
Taylor & Francis Group, an **informa** business

A BALKEMA BOOK

or near Suakin, Port Sudan, and Dungonab, (2) patch reefs and barrier reefs at Towartit and Wingate, (3) offshore reefs at Shaab Rumi and Sanganeb, and of course iconic man-made reefs· shipwrecks like the *Bahia Blanca*, better known as the *Umbria*, with davits breaking the surface on Wingate reef – its cargo holds still full of WW2 weaponry.

In addition to the data collected at the transect sites, dives on the Continental Shelf Station 2 (Precontinent II) garage at Shaab Rumi reef were made to map and measure coral colonies that

Figure 1.2 The platform of the Cambridge Coral Starfish Research Group was built on a reef that the group named "Harvey Reef" in the Towartit reef complex in the Red Sea. The aim was to investigate the causes of an apparent increase in aggregations of the coral-eating crown-of-thorns starfish. By making our temporary home on the reef, we came to better understand the processes by which coral reefs maintain a balance and how easy it is to tip that balance.

© Vine

Figure 1.3 (*a, b*) First photographs of white-spotted pufferfish *Arothron hispidus* eating the coral predating starfish *Acanthaster planci*. These show how the pufferfish attacks the underside of the spiny starfish by turning it over before eating.

© Vine

Figure 1.4 The pufferfish (*A. hispidus*) feeds mainly by biting the tips of branching corals, but we found it eating crown-of-thorns starfish (*A. planci*). This individual was caged as part of an experiment to test the responses of its prey.

© Vine

Figure 1.5 The dome-shaped garage formed part of the structure of "Conshelf 2" (also known as Precontinent 2), which in 1964 demonstrated the technologies for living under-water for extended periods of time. The author measured the growth rates of corals and a succession of marine life that settled on the garage.

© Frank Schneider/Alamy

had settled on the dome-shaped structure. Some species in the garage exhibited growth rates that were (and still are) among the fastest on record, with *Acropora pharaonis* extending 39 cm in length in only ten months. Such table corals play a significant role in rebuilding reefs that have been hit by events such as storms, floods, bleaching, or the CoTS (*Acanthaster planci*).

This study paid special attention to some of the observed threats to coral reefs, such as crown-of-thorns starfish (also known as CoTS), toxic thin film sponges like *Terpios hoshinota*, fast growing tunicates, phase shifts from coral domination to antagonistic macroalgal domination, overfishing of reef herbivores and pyramid predators, bacterial infections, and sedimentation.

As mentioned, when I began to write the notes for this book over 50 years ago, it was already with a sense of premonition that the luxuriant coral reefs of the central Red Sea were in a slow spiral of destruction. Scientists were drawing our attention to endangered species and vulnerable habitats – homes to vibrant populations of invertebrates, fish, marine reptiles, mammals, and seabirds – many of which were already on the IUCN's Red List. There was a general consensus that protection of species and conservation of coral reef habitats were priorities if we were ever to share these gifts of nature with future generations. We did not, however, anticipate that coral reefs themselves would be staring into the jaws of extinction within the coming decades – not in hundreds or thousands of years but in <u>this</u> century. Life in coral seas was changing, and it was changing fast. Along with the loss of reef-building corals, we were witnessing a widespread

Figure 1.6 Grey reef sharks (*Carcharhinus amblyrhynchos*) are among the most frequently seen sharks on central Red Sea reefs. Based on dietary analyses, they are not strictly speaking "apex predators" but should more accurately be considered as "high-level mesopredators." Under severe threat of illegal fishing, it remains unclear how their loss might affect coral reef ecosystems. It is an aggressive species, especially when provoked by divers. Attacks are generally preceded by threat displays.

© F. Jack Jackson/Alamy

phase shift – from corals to macroalgae – accompanied by the disappearance of many reef fish and their related flora and fauna.

Looming threats of environmental degradation, and the predicted impending death of many coral reefs, injected a sense of urgency into my studies. Overwhelmed by the biodiversity on Sudan's coral reefs (e.g., over 1,166 coral reef fishes) and wanting to capture this moment in time as a baseline against which to view these changes, I decided to focus on a few key elements. No two reefs were the same as each other, but all of the reefs shared patterns of responses to a common list of parameters that included: the physical and chemical characteristics of seawater; the structure of the reefs themselves, including the mechanisms by which they are built up or worn down; their communities of plants and animals; and their overall outlook. With these factors in mind, I set about recording dives on a wide range of reef types within a day's boat ride of Port Sudan.

Cousteau, Hass, and Cyril Crosssland were, with the giants who had come before them, pioneers of marine biological exploration (Hass 1952). SCUBA-equipment changed everything, and I now realise that our work was not only one of the *first* studies to detail the biodiversity and ecology of the coral world but also, in this Anthropocene of global warming, one of the *last* studies of its kind (Hass 1975).

Due to coral bleaching and other factors, huge areas of coral reefs can die in a single year. While there will be more refined work in the coming decades, analysing the forces at play in this battle for the survival of coral reefs, the almost inevitable rise in CO_2 levels, with its attendant acidification of the oceans, could turn such work into an obituary for Red Sea corals and their magnificent reefs – a far cry from the celebrations of those dives we made just 50 years ago.

Monitoring coral mortality is a depressing business that has driven a number of coral scientists to tears. When a whole reef turns white through the loss of symbiotic algae, the evidence of habitat breakdown is irrefutable. The scale of destruction can be local, regional, national, or international. The environmental impacts are almost impossible to overestimate, and the economic consequences are vast and unfathomable. It is a global problem that calls for a global response, but first we must understand the issues involved.

We dove into historic waters. "Roman Reef," better known as Shaab Rumi, was the chosen location for an experiment in which a new breed of divers moved in with the fish, not just for an hour or so but for days at a time. In the past, decompression tables placed strict limits on dive times. Now the tables were set aside, and the divers just stayed down, allowing their blood to become saturated with a mixture of oxygen and helium.

Figure 1.7 A typical rubble mound formed by skeletal remains of mostly *Acropora* staghorn corals. This was the aftermath of a crown-of-thorns starfish outbreak (known as CoTS or *Acanthaster planci*). Such loosely bound rubble may either destabilise the substrate and become coated by algae, hindering reef restoration, or become consolidated by calcification and crustose coralline algae (CCA), creating new surfaces for fresh corals to grow and flourish once again.

Discussing imaginative notions of underwater cities, Cousteau embraced the prospects that these discoveries presented, launching his Precontinent II (or Conshelf 2) expedition in 1963 and, in the process, making the world's first underwater campsite. Five divers lived for 30 days at 31 feet (9.4 m), and two divers spent a week further down the reef slope at 85 feet (25.9 m). In addition to having a pivotal role in submarine engineering, Cousteau's garage, as the docking station became known, provided a living laboratory to study the settlement, growth, and survival of reef-building corals under a range of conditions.

Figure 1.8 Red Sea adventures of Dr. Hans Hass and his secretary, to become wife, Lottie Berl taken 25 February 1953, by the *Sydney Morning Herald*. Hans Hass, along with Jacques Cousteau, was among the early pioneers of SCUBA diving.

There was so much to discover among these enticing waters. Berumen et al. (2013) provided an update on the general status of research on coral reef ecology of the Red Sea, which has long been recognised as a region of high biodiversity and endemism, commenting that our understanding of the ecology of coral reefs in the Red Sea has "lagged behind that of other coral reef systems." Their quantitative assessment of research published on the Red Sea, under eight specific topics, includes information on apex predators, connectivity, coral bleaching, coral reproductive biology, herbivory, marine protected areas (MPAs), non-coral invertebrates, and reef-associated bacteria. Comparing their results with those of other areas of major coral reef development, the authors note that the Red Sea had 1/6th the amount of research compared to Australia's Great Barrier Reef (GBR) and about 1/8th the amount of research in the Caribbean. These figures are distorted further by the fact that more than 50 percent of the Red Sea studies in the review took place in the Gulf of Aqaba, which represents less than 2 percent of the Red Sea area.

Explaining the significance of these figures, Berumen et al. (2013) observe that "data that could inform science-based management approaches are badly lacking in most Red Sea countries." Notwithstanding this relative lack of research, high levels of diversity and endemism (Ormond & Edwards 1987) are key features of Sudanese marine habitats (Head 1987) and are of major importance to an understanding of both the threats and the potential solutions to reef management and conservation in the region.

The Red Sea provides a window into predicting and managing the impact of climate change. As one of the world's most biodiverse coral reef regions (Stehli & Wells 1971; Polunin 1990), it has a significant role to play in the socioeconomic development of some of the most challenging biospheres on Earth. Cziesielski et al. (2021) wrote of this in a multi-author study entitled "Investing in Blue Natural Capital to Secure a Future for the Red Sea Ecosystems."

Chapter 2

Hydrography

An Ocean in the Making

One thing that makes the Sudanese Red Sea so extraordinary is its hydrography (Elsheikh et al. 2018). During winter, strong south-southeasterly winds prevail over the southern Red Sea and drive a surface inflow of relatively cold and nutrient-rich water from the Gulf of Aden into the Red Sea. During summer, north-northwesterly winds tend to extend over much of the Red Sea and may drive more saline surface water in the northern Red Sea to the south. Notably, monsoon winds are responsible for the subsurface influx of relatively cold and fresh Gulf of Aden Intermediate Water (GAIW), which enters the Red Sea through the Bab al-Mandab Strait during the summer (Kumagai et al. 2018).

A study of chemical and physical aspects of waters, together with zooplankton, around Port Sudan Harbour was published (Ahmed 2015) in 2015 in the *Journal of Marine Biology and Oceanography*).

A summary of the physical oceanography of the Red Sea as a whole has been recently published (Manasrah et al. 2019, 2020). The introduction to Rasul et al. (2019) contains a summary of the Red Sea's oceanography, while dedicated chapters deal with tides (Pugh et al. 2019), coastal lagoons, and other studies.

It is clear from this work that the distribution of fish species in Sudanese waters is related, at least in part, to temperatures and their effects on algal food sources. Indeed, seasonality plays a significant role in community composition in the Red Sea, particularly in the north and south, where extremes of sea surface temperature (SST) ranges are at their greatest.

Roth et al. (2021) conducted experiments on the lethality of turf algae at Saudi Arabia's Abu Shosha reef from January 2017 until January 2018. They showed that coral-dominated communities "exhibited 30% lower net productivity and 10 times higher calcification than algae-dominated communities." Net productivity and calcification were negatively correlated with temperature, whereas carbon losses via respiration and dissolved organic carbon release more than doubled at higher temperatures. In contrast, "algae-dominated communities doubled net productivity in summer, while calcification and dissolved organic carbon fluxes were unaffected." Higher temperatures seem to amplify the shift towards algal domination, even below coral bleaching thresholds, indicating that ocean warming may not only cause but also amplify coral–algal phase shifts in coral reefs.

Ateweberhan et al. (2016) describe the seasonality of corals and algae on reefs off the Eritrean coastline, drawing attention to the spatial and temporal variation in the biomass of four functional groups of coral reef algae. Canopy and foliose algae dominated the reef flat in winter, while crustose coralline algae (CCA) accounted for most of the macroalgal biomass throughout the year. Turf algae contributed the least to the total biomass in all reef zones. Biomass peaks shifted from midsummer on the inner reef flat to winter in the deeper zones. As temperatures rose, biomass decreased except in the case of turf algae on the shallow reef flat.

DOI: 10.1201/9781003335795-2

Figure 2.1 Traditional wooden dhows depended on seasonal winds and currents to trade and harvest the Red Sea's waters. Sailors, boat builders, and fishermen practiced skills that had hardly changed for centuries. Modern transport, coastal developments, and diving techniques wrought huge changes to the marine environment, and climate change has transformed our world.

© Vine

Chapter 3

Reef Conservation

Corals, almost by definition, are strong competitors for space on the reef, with their success depending on local conditions. Where corals and crustose coralline algae (CCA) are in direct competition, corals usually predominate. They are also superior competitors against turf algae, but anthropogenic influences that place corals under stressful conditions, may create favourable conditions for algae to grow. The presence or absence of herbivorous fish that browse turf algae, or a variety of other filamentous algae also influences the outcomes of these coral–algal "turf wars" (Abecasis 2013).

We have come a long way in terms of decoding the complex web of factors that spell success or failure on coral reefs: from field observations of individual coral colonies, such as those described in my own studies in Seychelles (Vine 1972b), to digitisation and AI simulation of impacts of altering currents, salinities, temperatures, fresh water, sedimentation, herbivory, and manipulation of local conditions. Coral reef simulators in Israel, the USA, Australia, and even Ireland host experiments that are revealing detailed information on individual species, and how they interact with one another under different physical regimes (Lang & Chornesky 1990).

The impact of certain herbivorous fish is now well known, with studies focussing on what happens if fish are removed from patches of reefs or settlement tiles. The classic example is the damselfish *Stegastes nigricans*, which creates nearly monocultural algae farms, as described by Hiroki Hata (2002), the present author (Vine 1974), and others. Within their defended territories, *S. nigricans* selectively weeds out indigestible algae, leaving just monocultural algae.

Mark Vermeij and colleagues have published extensively on the ecology of coral reefs and particularly the competition between corals and other species that are a constant feature of the reef biotope. He describes how marine algae, often belonging to invasive species, thrive where coral feeders are removed.

Reefs at Risk Revisited is a project of the World Resources Institute (WRI), developed and implemented in close collaboration with The Nature Conservancy (TNC), the World Fish Center, the International Coral Reef Action Network (ICRAN), the United Nations Environment Programme World Conservation Monitoring Centre (UNEPWCMC), and the Global Coral Reef Monitoring Network (GCRMN) together with many other government agencies, international organisations, research institutions, universities, and nongovernmental bodies. The preamble of its report reminds us of the importance of coral reefs to our own life on earth: "coral reefs, the "rain forests of the sea," are among the most biologically rich and productive ecosystems on earth." Such reefs also provide valuable bioresources to millions of coastal dwellers, including food, tourism attractions, and natural defences against storms (Lieske & Myers 2004).

The outlook for coral reefs has been dramatically altered by global warming and associated coral bleaching in which corals lose their symbiotic algae exposing their white skeletons. As

DOI: 10.1201/9781003335795-3

Figure 3.1 Scalefin anthias (*Pseudanthias squamipinnis*) are common on healthy fringing and patch reefs to 40 m or so, often forming large aggregations. Males are territorial and maintain a harem of females. After removal of a male from its harem, it takes about one to two weeks for the ranking female to change sex.

© Hans Sjöholm

Figure 3.2 A diver passes by a very large fan coral near Abu Galawa reef, Yanbu, Saudi Arabia.

© Hans Sjöholm

Figure 3.3 Herbivorous fish such as these tangs are important in the rehabilitation of the reef.
© Hans Sjöholm

oceans warm, the extent of coral bleaching is becoming more widespread and damaging to healthy reefs. Furthermore, carbon dioxide levels are rising in the atmosphere as well as the oceans, rendering our seas more acidic.

It was increasingly clear that the diversity of species and the high degree of endemicity and connectivity of ecosystems needed to be taken into account in planning future conservation programmes. Meanwhile, genetic analysis of some specimens revealed the existence of crypto-species, whose status as separate species could be confirmed only through DNA sequencing. This has been recently explored in relation to the genomes of nudibranchs found in the Red Sea (Osman 2016).

Gajdzik et al. (2021) propose a "portfolio of climate-tailored approaches to advance the design of marine protected areas (MPAs) in the Red Sea." Commenting on the apparent lack of effective management in protected marine areas, including the UNESCO World Heritage Sites at Sanganeb and Dungonab, they conclude that all too often, the planning of MPAs does not take into account rapidly changing climatic conditions. The solution, they claim, is to "tailor the design of MPAs by integrating approaches to enhance climate change mitigation and adaptation."

To illustrate how this approach might work, Gajdzik et al. (2021) take the example of coral bleaching and the aim of protecting reefs from different thermal regimes, thus reducing the risk that all protected reefs will bleach simultaneously. They also point to the importance of preserving genetic (Miller & Ayre 2008) connectivity patterns that are assisted by mesoscale eddies and mangrove forests that act as major carbon sinks. The results of this more bespoke approach

Figure 3.4 Coral reefs support some of the highest levels of biodiversity on our planet. Colour and aesthetics are attractive features of their natural ecosystems. This Amara Spanish dancer *Hexabranchus* was photographed off Yemen in the southern Red Sea.

© Vine

to conserving Red Sea reefs could transform the MPA selection process into one that prioritises key habitats with the aim of optimising regional outcomes.

Abelson (2020) asked the question, "Are we sacrificing the future of coral reefs on the altar of the 'climate change' narrative?" He had a good point, one whose importance is emphasised by the extent of obfuscation increasingly apparent in national statements dealing with efforts to address the climate change issue at, for example, speeches at the COP26 international climate conference that took place in Glasgow from 31 October to 12 November 2021.

Recent publications, citing the widely held negative prognosis for coral reefs, have regarded climate change as the key factor that needs to be addressed (Kleinhaus et al. 2020).

This is partially justified, but given the scale of devastation that coral bleaching can wreak, it is not the whole picture. Collapse of local fish stocks, increases in agricultural and industrial effluents, siltation, and chemical pollution are all stressors that can trigger reef mortalities. There is little point in solving the issues directly connected to climate change if we don't simultaneously resuscitate coastal marine habitats, whose health is essential for the reefs to survive and flourish (Camp et al. 2018). Is this feasible? Will at least some reef-building corals adapt to survive the environmental changes taking place on the reefs? Will genetic variations emerge that are capable of making this happen? Extreme conditions, such as those occurring at CO_2 vent sites, may offer a limited view of the future and provisional analogues for predicted coral environments.

Effective management of recuperative actions on coral reefs should include, but not necessarily be limited to:

1   Development of coral farms to produce seed stock for replanting and growing-out of certain coral species
2   Consideration of temporary shade structures that help to protect young, cultivated coral colonies
3   Introduction of banned fishing and diving zones at sites of previous coral reefs that have deteriorated due to various causes
4   Avoidance of sediment-laden run-off
5   Introduction and enforcement of conservation legislation
6   Implementation of awareness and education programmes
7   Encouragement of citizen science as a way of focussing more eyes and ears on changes on reefs
8   Research and publication of factors affecting the growth and decay of coral reefs
9   Removal of polluting influences including cooling water
10 Sharing experiences with specialist scientists in other countries.

To such a checklist of interventions, I would also add sustainable monetisation of coral reefs so that it is economically attractive to conserve coral reefs on an ongoing basis. Examples are to be found in the pharmaceutical field and that of bone implants as well as molecular studies.

Coral propagation in the form of fragmentation, husbandry, and out planting of branching coral species (especially table *Acropora* species), aimed at providing a stock of corals for replanting on reefs, has shown promising results.

Section I

Fish

Figure I.1 Nemo Garden near Yanbu, Saudi Arabia. Species include soft coral (*Dendroneph-thya hemprichi*), Red Sea anemonefish (*Amphiprion bicinctus*), three-spot dascyllus (*Dascyllus trimaculatus*), scalefin anthias (*Pseudanthias squamipinnis*), and magnificent anemones (*Heteractis magnifica*).

©Hans Sjöholm

One of the most impressive features of diving over Sudan's reefs is its extraordinary variety of fish. Summarising studies on taxonomy and zoogeography of Red Sea fishes, Bogorodsky and Randall (2019) recognised 1,166 species known from the Red Sea, almost 15 percent of which they regarded as endemic. Despite some detailed studies, information on the distribution of Red Sea species is far from complete, and new species are still being discovered (e.g., Uiblein 2021).

DOI: 10.1201/9781003335795-4

Taxonomic studies laid the groundwork for more behavioural studies. In one example involving studies by Edwards and Rosewell (1981) on reefs near Port Sudan, the authors recommend that 68 species of fish be included in local checklists on the basis of their abundance and ease of identification. Compare this number with a similar list of 76 species proposed by Vine and Vine (1980). The latter recorded 204 fish species during their surveys.

Fish surveys are important in terms of understanding the reef fish assemblages upon which healthy reef communities depend. In general terms, the greater the diversity of associated species present, the more resilient is the overall community compared to the vulnerability of low-diversity habitats that are dominated by just a few species and therefore susceptible to population crashes. Nanami and Nishihira (2003) point out that stability in reef fish communities depends in part on whether distribution is isolated or continuous. The latter tend to be more stable (Nanami & Nishihira 2003).

Robitzch and Berumen (2020) studied the breeding behaviour of Red Sea coral reef fishes, pointing out that while the norm is for recruitment to peak during summer months, this tends not to be the case in the high temperatures of the Sudanese Red Sea, where recruitment and plankton concentrations peak in winter.

Chapter 4

Fish Surveys

A number of studies on fish distribution have taken place on accessible reefs on the Sudanese coastline and offshore reefs. Edwards and Rosewell (1981) studied the distribution of 68 fish species on Harvey Reef, partitioning the reef environment as (A) confined to reef top, (B) predominantly reef top, (C) confined to reef crest, (D) reef crest and shallow slope (uncommon below 10–15 m), (E) reef crest and reef slope, (F) reef slope, (G) deeper reef slope (rare above 10 m depth), or (H) ubiquitous (reef flat to 20 m depth).

Figure 4.1 Blue-cheeked or masked butterflyfish (*Chaetodon semilarvatus*) are often seen at depths less than 20 m and are associated with healthy coral. They typically swim in pairs or small aggregations hovering beneath *Acropora* tables. They feed on hard and soft corals in the late afternoon.

© Hans Sjöholm

DOI: 10.1201/9781003335795-5

Figure 4.2 Yellowbar angelfish (*Pomacanthus maculosus*) prefer sheltered locations where they feed on sponges and algae.

© Vine

They concluded that: "the reef flat supports large numbers of relatively few species of fish and with the increasing exposure of the reefs, the non-cryptic fish communities of the reef flat become apparently less diverse." Meanwhile, in deeper water (15–20 m), there is greater species diversity than on the reef flat, but the fish population is considerably less than that on the reef flat, presumably as a result of lower primary productivity (cf. Vine 1974).

Kattan (2017) studied fishing pressure on fish in the Sudanese Red Sea, pointing to the still high biomass and diversity levels of Sudan's reefs compared to those off Saudi Arabia.

Anna Knochel's thesis included a study of manta rays in the Sudanese Red Sea, following the discovery of a large reef manta ray (*Mobula alfredi*) aggregation that had been observed off the north Sudanese Red Sea coast since the 1950s. SPOT 5 tags were secured to three manta rays.

Figure 4.3 Feeding aggregations of *Manta alfredi* occurred near Mesharifa island, south of Dungonab Bay. A closely related, or possibly synonymous species is *Manta birostris*.

© Vine

Key locations were reported to occur within 15 km of a proposed large-scale island development. These are important from a conservation viewpoint (Kessel et al. 2017).

The checklist of fishes recorded during the reef site dives in 1974–1975 provides a baseline for comparing with today's fish populations.

Chapter 5

Fish Recorded at Reef Sites

The importance of deeper water populations of coral reef fish has been recognised by several researchers who have focused on their roles as refuges for threatened species in degraded habitats (MacDonald 2016). Srinivasan (2003) writes on the topic in "Depth distributions of coral reef fishes: the influence of microhabitat structure, settlement, and post-settlement processes."

In my own studies, the group of species with the greatest maximum depth range generally had the highest percentage occurrence on reef sites investigated. Groups with smaller vertical distribution ranges had lower percentage occurrences at investigated reef sites. The tentative conclusion drawn is that species with large vertical distribution depth ranges are less particular

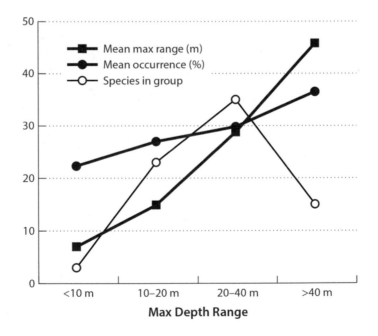

Figure 5.1 Correlation between maximum depth range and percentage occurrence.
© Vine

DOI: 10.1201/9781003335795-6

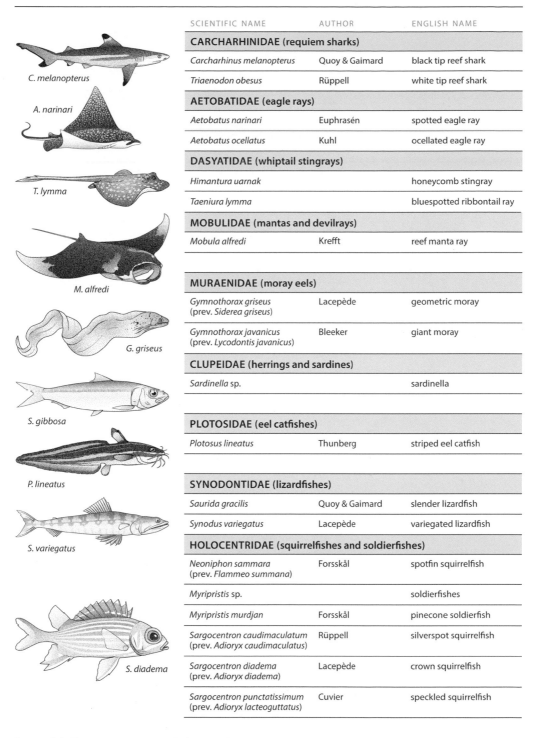

SCIENTIFIC NAME	AUTHOR	ENGLISH NAME
CARCHARHINIDAE (requiem sharks)		
Carcharhinus melanopterus	Quoy & Gaimard	black tip reef shark
Triaenodon obesus	Rüppell	white tip reef shark
AETOBATIDAE (eagle rays)		
Aetobatus narinari	Euphrasén	spotted eagle ray
Aetobatus ocellatus	Kuhl	ocellated eagle ray
DASYATIDAE (whiptail stingrays)		
Himantura uarnak		honeycomb stingray
Taeniura lymma		bluespotted ribbontail ray
MOBULIDAE (mantas and devilrays)		
Mobula alfredi	Krefft	reef manta ray
MURAENIDAE (moray eels)		
Gymnothorax griseus (prev. *Siderea griseus*)	Lacepède	geometric moray
Gymnothorax javanicus (prev. *Lycodontis javanicus*)	Bleeker	giant moray
CLUPEIDAE (herrings and sardines)		
Sardinella sp.		sardinella
PLOTOSIDAE (eel catfishes)		
Plotosus lineatus	Thunberg	striped eel catfish
SYNODONTIDAE (lizardfishes)		
Saurida gracilis	Quoy & Gaimard	slender lizardfish
Synodus variegatus	Lacepède	variegated lizardfish
HOLOCENTRIDAE (squirrelfishes and soldierfishes)		
Neoniphon sammara (prev. *Flammeo summana*)	Forsskål	spotfin squirrelfish
Myripristis sp.		soldierfishes
Myripristis murdjan	Forsskål	pinecone soldierfish
Sargocentron caudimaculatum (prev. *Adioryx caudimaculatus*)	Rüppell	silverspot squirrelfish
Sargocentron diadema (prev. *Adioryx diadema*)	Lacepède	crown squirrelfish
Sargocentron punctatissimum (prev. *Adioryx lacteoguttatus*)	Cuvier	speckled squirrelfish

Figure 5.2 Fish species recorded at reef sites. Illustrations by Fiona Martin (Continued).
© Vine

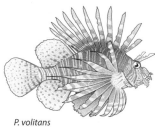

N. savayensis

D. excisus

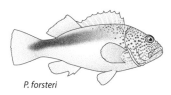

P. volitans

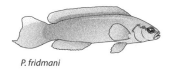

E. merra

P. forsteri

P. fridmani

SCIENTIFIC NAME	AUTHOR	ENGLISH NAME
APOGONIDAE (cardinalfishes)		
Apogon exostigma	Jordan & Starks	eyeshadow cardinalfish
Apogon semiornatus	Peters	oblique-banded cardinalfish
Cheilodipterus quinquelineatus	Cuvier & Valenciennes	fiveline cardinalfish
Nectamia savayensis	Günther	Samoan cardinalfish
SYNGNATHIDAE (pipefishes and seahorses)		
Corythoichthys flavofasciatus	Rüppell	network pipefish
Doryrhamphus excisus (synonym *Doryrhamphus melanopleura*)	Kaup	bluestripe pipefish
SCORPAENIDAE (scorpionfishes)		
Pterois radiata	Cuvier & Valenciennes	clearfin lionfish
Pterois volitans	Linnaeus	red lionfish
SERRANIDAE (groupers and allies)		
Aethaloperca rogaa	Forsskål	redmouth grouper
Anyperodon leucogrammicus	Cuvier & Valenciennes	slender grouper
Cephalopholis argus	Bloch & Schneider	peacock grouper
Cephalopholis hemistiktos	Rüppell	halfspotted grouper
Cephalopholis miniata	Forsskål	coral grouper
Cephalopholis sexmaculata	Rüppell	sixspot grouper
Diploprion drachi	Roux-Esteve	yellowface soapfish
Epinephelus areolatus	Forsskål	aerolate grouper
Epinephelus chlorostigma	Valenciennes	brownspotted grouper
Epinephelus merra	Bloch	honeycomb grouper
Epinephelus summana	Forsskål	summana grouper
Epinephelus tauvina	Forsskål	greasy grouper
Pseudanthias squamipinnis	Peters	sea goldie, scalefin anthias
Variola louti	Forsskål	lunartail grouper
CIRRHITIDAE (hawkfishes)		
Cirrhitichthys oxycephalus	Bleeker	pixy hawkfish
Cirrhitus pinnulatus	Bloch & Schneider	stocky hawkfish
Oxycirrhites typus	Bleeker	longnose hawkfish
Paracirrhites forsteri	Bloch & Schneider	blackside hawkfish
PSEUDOCHROMIDAE (dottybacks and snakelets)		
Pseudochromis flavivertex	Rüppell	sunrise dottyback
Pseudochromis fridmani	Klausewitz	orchid dottyback

Figure 5.2 (Continued)

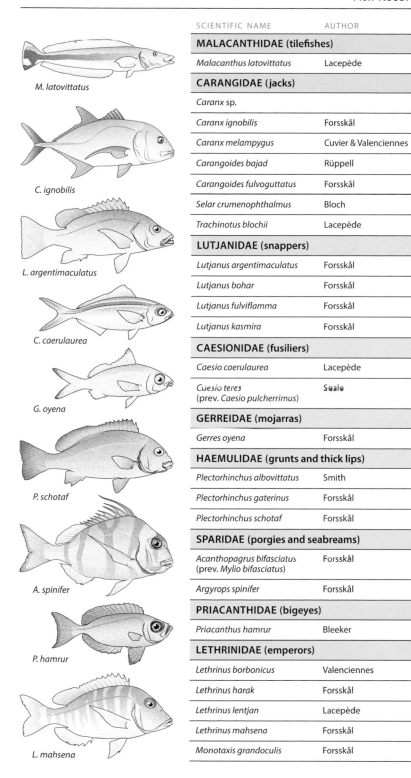

SCIENTIFIC NAME	AUTHOR	ENGLISH NAME
MALACANTHIDAE (tilefishes)		
Malacanthus latovittatus	Lacepède	striped blanquillo
CARANGIDAE (jacks)		
Caranx sp.		jacks, trevallies, kingfishes
Caranx ignobilis	Forsskål	giant trevally
Caranx melampygus	Cuvier & Valenciennes	bluefin trevally
Carangoides bajad	Rüppell	orangespotted jack
Carangoides fulvoguttatus	Forsskål	yellowspotted jack
Selar crumenophthalmus	Bloch	bigeye scad
Trachinotus blochii	Lacepède	snubnose dart
LUTJANIDAE (snappers)		
Lutjanus argentimaculatus	Forsskål	mangrove red snapper
Lutjanus bohar	Forsskål	two-spot red snapper
Lutjanus fulviflamma	Forsskål	dory snapper
Lutjanus kasmira	Forsskål	blue-striped snapper
CAESIONIDAE (fusiliers)		
Caesio caerulaurea	Lacepède	blue and gold fusilier
Caesio teres (prev. *Caesio pulcherrimus*)	Seale	yellow and blueback fusilier
GERREIDAE (mojarras)		
Gerres oyena	Forsskål	common silver-biddy
HAEMULIDAE (grunts and thick lips)		
Plectorhinchus albovittatus	Smith	two-striped sweetlips
Plectorhinchus gaterinus	Forsskål	blackspotted rubberlip
Plectorhinchus schotaf	Forsskål	minstrel sweetlips
SPARIDAE (porgies and seabreams)		
Acanthopagrus bifasciatus (prev. *Mylio bifasciatus*)	Forsskål	doublebar bream
Argyrops spinifer	Forsskål	king soldier bream
PRIACANTHIDAE (bigeyes)		
Priacanthus hamrur	Bleeker	goggle-eye
LETHRINIDAE (emperors)		
Lethrinus borbonicus	Valenciennes	snubnose emperor
Lethrinus harak	Forsskål	thumbprint emperor
Lethrinus lentjan	Lacepède	pink ear emperor
Lethrinus mahsena	Forsskål	sky emperor
Monotaxis grandoculis	Forsskål	bigeye emperor

Figure 5.2 (Continued)

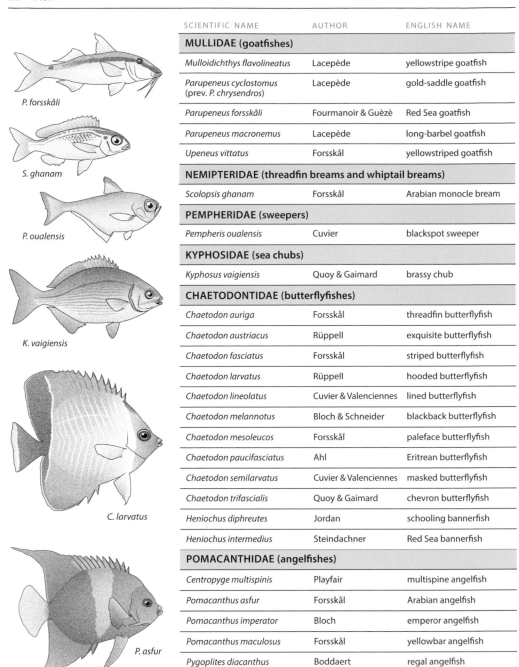

SCIENTIFIC NAME	AUTHOR	ENGLISH NAME
MULLIDAE (goatfishes)		
Mulloidichthys flavolineatus	Lacepède	yellowstripe goatfish
Parupeneus cyclostomus (prev. *P. chrysendros*)	Lacepède	gold-saddle goatfish
Parupeneus forsskåli	Fourmanoir & Guèzè	Red Sea goatfish
Parupeneus macronemus	Lacepède	long-barbel goatfish
Upeneus vittatus	Forsskål	yellowstriped goatfish
NEMIPTERIDAE (threadfin breams and whiptail breams)		
Scolopsis ghanam	Forsskål	Arabian monocle bream
PEMPHERIDAE (sweepers)		
Pempheris oualensis	Cuvier	blackspot sweeper
KYPHOSIDAE (sea chubs)		
Kyphosus vaigiensis	Quoy & Gaimard	brassy chub
CHAETODONTIDAE (butterflyfishes)		
Chaetodon auriga	Forsskål	threadfin butterflyfish
Chaetodon austriacus	Rüppell	exquisite butterflyfish
Chaetodon fasciatus	Forsskål	striped butterflyfish
Chaetodon larvatus	Rüppell	hooded butterflyfish
Chaetodon lineolatus	Cuvier & Valenciennes	lined butterflyfish
Chaetodon melannotus	Bloch & Schneider	blackback butterflyfish
Chaetodon mesoleucos	Forsskål	paleface butterflyfish
Chaetodon paucifasciatus	Ahl	Eritrean butterflyfish
Chaetodon semilarvatus	Cuvier & Valenciennes	masked butterflyfish
Chaetodon trifascialis	Quoy & Gaimard	chevron butterflyfish
Heniochus diphreutes	Jordan	schooling bannerfish
Heniochus intermedius	Steindachner	Red Sea bannerfish
POMACANTHIDAE (angelfishes)		
Centropyge multispinis	Playfair	multispine angelfish
Pomacanthus asfur	Forsskål	Arabian angelfish
Pomacanthus imperator	Bloch	emperor angelfish
Pomacanthus maculosus	Forsskål	yellowbar angelfish
Pygoplites diacanthus	Boddaert	regal angelfish

P. forsskåli

S. ghanam

P. oualensis

K. vaigiensis

C. larvatus

P. asfur

Figure 5.2 (Continued)

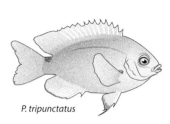

P. tripunctatus

SCIENTIFIC NAME	AUTHOR	ENGLISH NAME
POMACENTRIDAE (damselfishes)		
Abudefduf saxatilis	Linnaeus	sergeant-major
Abudefduf sexfasciatus	Lacepède	scissortail sergeant
Abudefduf sordidus	Forsskål	blackspot sergeant
Abudefduf vaigiensis	Quoy & Gaimard	Indo-Pacific sergeant
Amblyglyphidodon sp.		damselfishes
Amblyglyphidodon leucogaster	Bleeker	yellowbelly damselfish
Amphiprion bicinctus	Rüppell	twoband anemonefish
Chromis caerulea	Cuvier	green chromis
Chromis ternatensis	Bleeker	ternate chromis
Dascyllus aruanus	Linnaeus	whitetail dascyllus
Dascyllus marginatus	Rüppell	Red Sea dascyllus
Dascyllus trimaculatus	Rüppell	threespot dascyllus
Neoglyphidodon melas	Cuvier	bowtie damselfish
Neopomacentrus sp.		lyretail damselfishes
Pomacentrus albicaudatus	Baschieri-Salvadori	whitefin damsel
Pomacentrus sulfureus	Klunzinger	sulphur damsel
Pomacentrus tripunctatus	Cuvier	threespot damsel
Pycnochromis dimidiatus	Klunzinger	chocolatedip chromis
Stegastes lacrymatus	Quoy & Gaimard	white spotted devil
Stegastes lividus	Bloch & Schneider	blunt snout gregory
Stegastes nigricans	Lacepède	dusky farmerfish
LABRIDAE (wrasses)		
Bodianus anthioides	Bennett	lyretail hogfish
Bodianus axillaris	Quoy & Gaimard	axilspot hogfish
Cheilinus fasciatus	Bloch	redbreasted wrasse
Cheilinus lunulatus	Forsskål	broomtail wrasse
Cheilinus trilobatus	Lacepède	tripletail wrasse
Cheilinus undulatus	Rüppell	humphead wrasse
Coris caudimaculata	Quoy & Gaimard	spottail coris
Coris cuvieri (prev. *Coris gaimard africans*)	Smith	African coris
Coris variegata	Rüppell	dapple coris
Epibulus insidiator	Pallas	sling-jaw wrasse

Figure 5.2 (Continued)

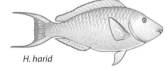

P. octotaenia

H. harid

E. midas

SCIENTIFIC NAME	AUTHOR	ENGLISH NAME
LABRIDAE (wrasses, cont.)		
Gomphosus caeruleus	Lacepède	green birdmouth wrasse
Halichoeres hortulanus (prev. Halichoeres centriquadrus)	Lacepède	checkerboard wrasse
Halichoeres scapularis	Bennett	zigzag wrasse
Hemigymnus fasciatus	Bloch	barred thicklip
Labroides dimidiatus	Valenciennes	bluestreak cleaner wrasse
Larabicus quadrilineatus	Rüppell	fourline wrasse
Novaculichthys taeniourus	Lacepède	rockmover wrasse
Oxycheilinus celebicus	Bleeker	celebes wrasse
Paracheilinus octotaenia	Formanoir	Red Sea eightline flasher
Pseudocheilinus hexataenia	Bleeker	sixline wrasse
Pteragogus pelycus	Randall	sideburn wrasse
Pterogogus sp.		wrasses
Stethojulis albovittata	Bonaterre	bluelined wrasse
Thalassoma lunare	Linnaeus	moon wrasse
Thalassoma purpureum	Forsskål	surge wrasse
Thalassoma rueppellii	Klunziger	Klunzinger's wrasse
SCARIDAE (parrotfishes)		
Bolbometopon muricatum	Valenciennes	green humphead parrotfish
Cetoscarus bicolor	Rüppell	bicolour parrotfish
Chlorurus gibbus (prev. Scarus gibbus)	Rüppell	heavybeak parrotfish
Chlorurus sordidus (prev. Scarus sordidus)	Forsskål	daisy parrotfish
Hipposcarus harid	Forsskål	candelamoa parrotfish
Scarus ferrugineus	Forsskål	rusty parrotfish
Scarus ghobban	Forsskål	blue-barred parrotfish
Scarus niger	Forsskål	dusky parrotfish
BLENNIIDAE (blennies)		
Aspidontus tractus	Fowler	mimic blenny
Ecsenius aroni	Springer	Aron's blenny
Ecsenius frontalis	Valenciennes	smooth-fin blenny
Ecsenius gravieri	Pellegrin	Red Sea mimic blenny
Ecsenius midas	Starck	Persian blenny
Istiblennius sp.		combtooth blennies

Figure 5.2 (Continued)

SCIENTIFIC NAME	AUTHOR	ENGLISH NAME
BLENNIIDAE (blennies, cont.)		
Meiacanthus nigrolineatus	Smith-Vaniz	blackline fangblenny
Mimoblennius cirrosus	Smith-Vaniz & Springer	fringed blenny
Petroscirtes sp.		combtooth blennies
Plagiotremus tapeinosoma	Bleeker	piano fangblenny
Plagiotremus townsendi	Regan	Townsend's fangblenny
GOBIIDAE (gobies)		
Acentrogobius sp.		gobies
Amblyeleotris steinitzi (prev. *Cryptocentrus steinitzi*)	Klausewitz	Steinitz' prawn-goby
Amblygobius albimaculatus	Rüppell	butterfly goby
Amblygobius sp. (cf. *decessatus*)		orange-striped goby
Cryptocentrus sp.		Watchman gobies
Cryptocentrus caeruleopunctatus	Rüppell	Harlequin prawn-goby
Cryptocentrus lutheri	Klausewitz	Luther's prawn-goby
Ctenogobiops maculosus	Fourmanoir	Seychelles shrimpgoby
Koumansetta hectori	Smith	Hector's goby
Lotilia graciliosa	Klausewitz	whitecap goby
EPHIPPIDAE (spadefishes)		
Platax orbicularis	Forsskål	orbicular batfish
SIGANIDAE (rabbitfishes)		
Siganus stellatus	Forsskål	brown-spotted spinefoot
ACANTHURIDAE (surgeonfishes)		
Acanthurus gahhm	Forsskål	black surgeonfish
Acanthurus nigrofuscus	Forsskål	brown surgeonfish
Acanthurus sohal	Forsskål	Sohal surgeonfish
Ctenochaetus striatus	Quoy & Gaimard	striated surgeonfish
Naso brevirostris	Cuvier	spotted unicornfish
Naso hexacanthus	Bleeker	sleek unicornfish
Naso lituratus (prev. *Callicanthus lituratus*)	Bloch & Schneider	orangespine unicornfish
Naso unicornis	Forsskål	bluespine unicornfish
Zebrasoma desjardinii	Bennett	Indian sail-fin surgeonfish
Zebrasoma xanthurum	Blyth	yellowtail tang

A. steinitzi

P. orbicularis

S. stellatus

A. sohal

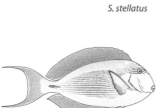

Figure 5.2 (Continued)

SCIENTIFIC NAME	AUTHOR	ENGLISH NAME
SPHYRAENIDAE (barracudas)		
Sphyraena barracuda	Walbaum	great barracuda
Sphyraena jello	Cuvier	pickhandle barracuda
BALISTIDAE (triggerfishes)		
Balistapus undulatus	Mungo-Park	orange-lined triggerfish
Balistoides viridescens	Bloch & Schneider	titan triggerfish
Odonus niger	Rüppell	red-toothed triggerfish
Pseudobalistes flavimarginatus	Rüppell	yellowmargin triggerfish
Rhinecanthus assasi	Forsskål	Picasso triggerfish
Sufflamen albicaudatum (prev. *Sufflamen albicaudatus*)	Rüppell	bluethroat triggerfish
Sufflamen chrysoptera (prev. *Hemibalistes chrysoptera*)	Bloch	halfmoon triggerfish
MONACANTHIDAE (filefishes)		
Cantherines pardalis (prev. *Cantherines pardas*)	Rüppell	honeycomb filefish
Canthigaster margaritata	Rüppell	pearl toby
Monacanthus sp.		filefishes
Oxymonacanthus halli	Marshall	Red Sea longnose filefish
OSTRACIIDAE (boxfishes and cowfishes)		
Ostracion cubicum (prev. *Ostracion argus*)	Rüppell	yellow boxfish
Ostracion cyanurus	Rüppell	bluetail trunkfish
TETRAODONTIDAE (puffers)		
Arothron hispidus	Lacepède	white-spotted puffer
Arothron nigropunctatus (*diadematus*)	Bloch & Schneider	blackspotted puffer
DIODONTIDAE (porcupinefishes)		
Diodon hystrix	Linnaeus	spot-fin porcupinefish

S. barracuda

O. niger

O. halli

O. cubicum

A. hispidus

D. hystrix

Figure 5.2 (Continued)

in habitat selection and are therefore more widespread on different reef sites. It is not, however, a fixed rule, and there are several important exceptions. For example, *Acanthurus sohal* has a narrow depth range (20 m) but a high percentage of occurrence (46 percent on reef sites). This is apparently because the species favours the shallow zone behind the reef crest and is a very widespread coloniser of this habitat.

Fish species recorded at reef sites are listed here, followed by visual comparisons of relative frequency of occurrence, distribution relative to depth, and distribution across reef sites (also referred to as transects).

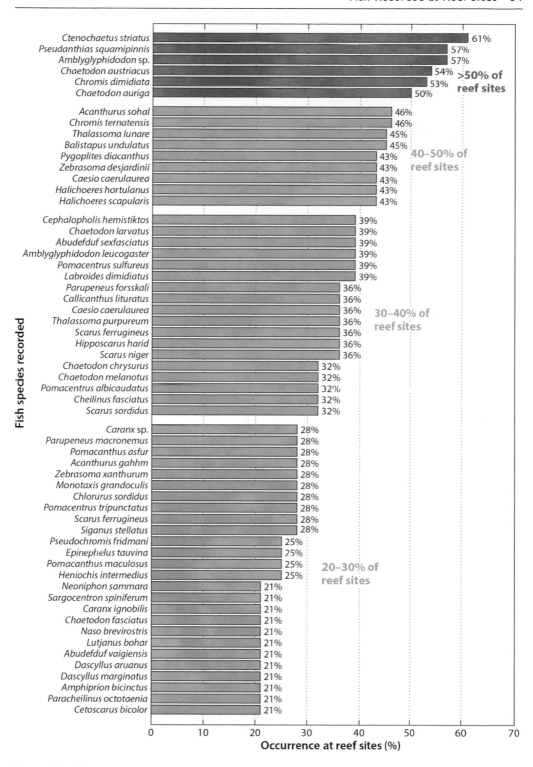

Figure 5.3 Relative frequency of occurrence of fish species.

© Vine

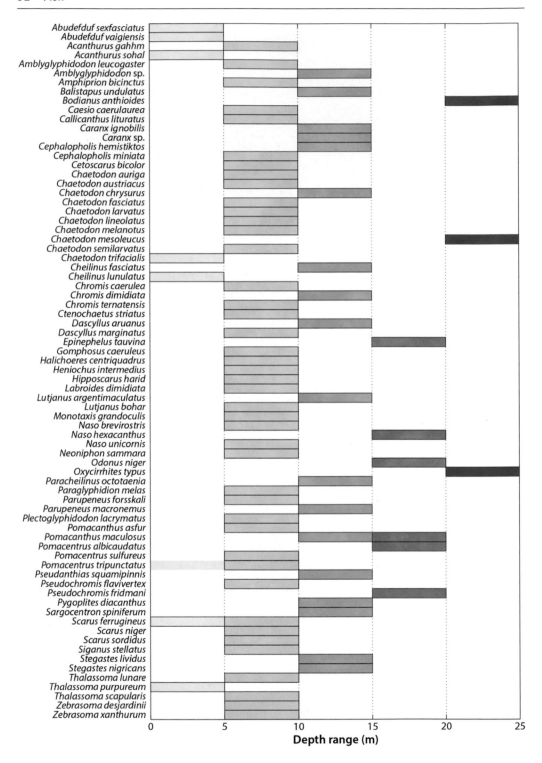

Figure 5.4 Distribution of fish species relative to depth.

© Vine

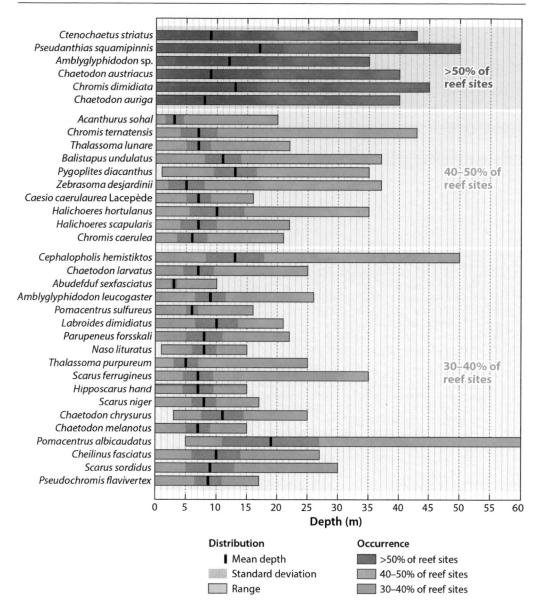

Figure 5.5 Distribution of fishes at transect sites (Continued).

© Vine

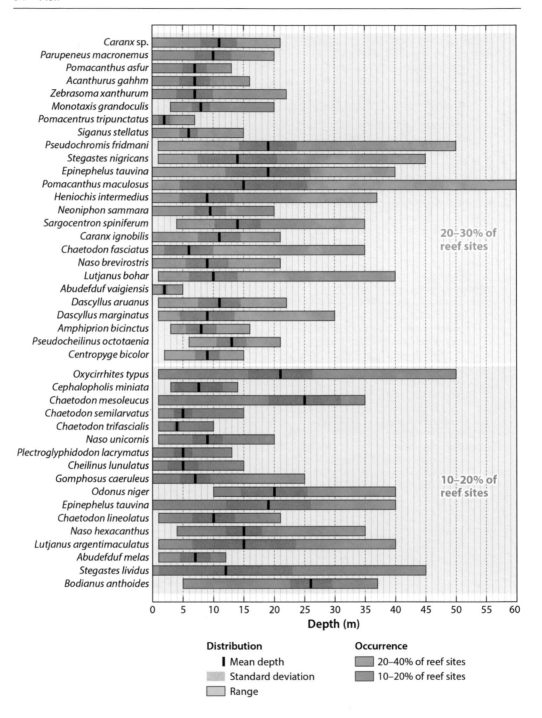

Figure 5.5 (Continued)

Chapter 6

Fish in Suakin Harbour

Fish recorded in the Suakin region are featured here. The list was compiled with the assistance of the late Dr. J.E. Randall.

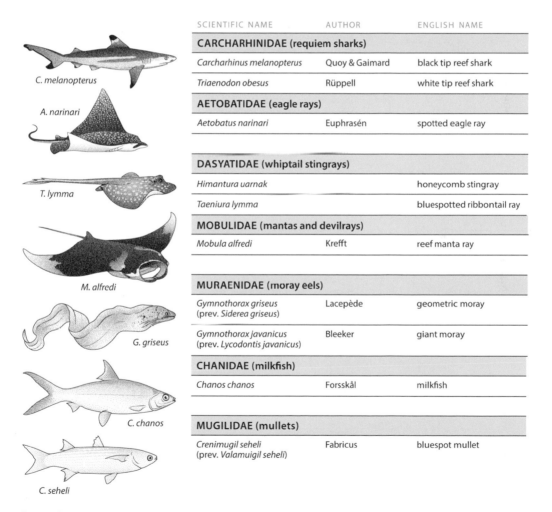

SCIENTIFIC NAME	AUTHOR	ENGLISH NAME
CARCHARHINIDAE (requiem sharks)		
Carcharhinus melanopterus	Quoy & Gaimard	black tip reef shark
Triaenodon obesus	Rüppell	white tip reef shark
AETOBATIDAE (eagle rays)		
Aetobatus narinari	Euphrasén	spotted eagle ray
DASYATIDAE (whiptail stingrays)		
Himantura uarnak		honeycomb stingray
Taeniura lymma		bluespotted ribbontail ray
MOBULIDAE (mantas and devilrays)		
Mobula alfredi	Krefft	reef manta ray
MURAENIDAE (moray eels)		
Gymnothorax griseus (prev. *Siderea griseus*)	Lacepède	geometric moray
Gymnothorax javanicus (prev. *Lycodontis javanicus*)	Bleeker	giant moray
CHANIDAE (milkfish)		
Chanos chanos	Forsskål	milkfish
MUGILIDAE (mullets)		
Crenimugil seheli (prev. *Valamuigil seheli*)	Fabricus	bluespot mullet

Figure 6.1 Fish recorded in Suakin Harbour (Continued). Illustrations by Fiona Martin. © Vine.

DOI: 10.1201/9781003335795-7

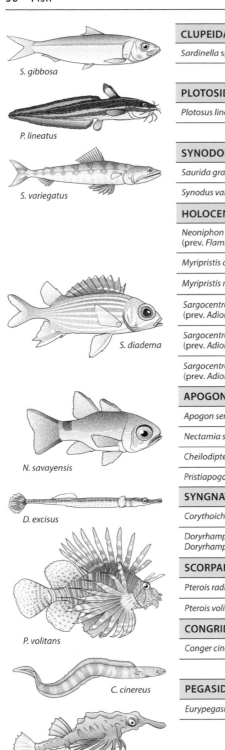

S. gibbosa

P. lineatus

S. variegatus

S. diadema

N. savayensis

D. excisus

P. volitans

C. cinereus

E. draconis

CLUPEIDAE (herrings and sardines)		
Sardinella sp.		sardinella

PLOTOSIDAE (eel catfishes)		
Plotosus lineatus	Thunberg	striped eel catfish

SYNODONTIDAE (lizardfishes)		
Saurida gracilis	Quoy & Gaimard	slender lizardfish
Synodus variegatus	Lacepède	variegated lizardfish

HOLOCENTRIDAE (squirrelfishes and soldierfishes)		
Neoniphon sammara (prev. Flammeo summana)	Forsskål	spotfin squirrelfish
Myripristis chryseres	Jordan & Evermann	yellowfin soldierfish
Myripristis murdjan	Forsskål	pinecone soldierfish
Sargocentron caudimaculatum (prev. Adioryx caudimaculatus)	Rüppell	silverspot squirrelfish
Sargocentron diadema (prev. Adioryx diadema)	Lacepède	crown squirrelfish
Sargocentron punctatissimum (prev. Adioryx lacteoguttatus)	Cuvier	speckled squirrelfish

APOGONIDAE (cardinalfishes)		
Apogon semiornatus	Peters	oblique-banded cardinalfish
Nectamia savayensis	Günther	Samoan cardinalfish
Cheilodipterus quinquelineatus	Cuvier & Valenciennes	fiveline cardinalfish
Pristiapogon exostigma	Jordan & Starks	narrowstripe cardinalfish

SYNGNATHIDAE (pipefishes and seahorses)		
Corythoichthys intestinalis	Ramsay	scribbled pipefish
Doryrhamphus excisus (synonym Doryrhamphus melanopleura)	Kaup	bluestripe pipefish

SCORPAENIDAE (scorpionfishes)		
Pterois radiata	Cuvier & Valenciennes	clearfin lionfish
Pterois volitans	Linnaeus	red lionfish

CONGRIDAE (conger and garden eels)		
Conger cinereus	Rüppell	longfin African conger

PEGASIDAE (seamoths)		
Eurypegasus draconis	Linnaeus	short dragonfish

Figure 6.1 (Continued)

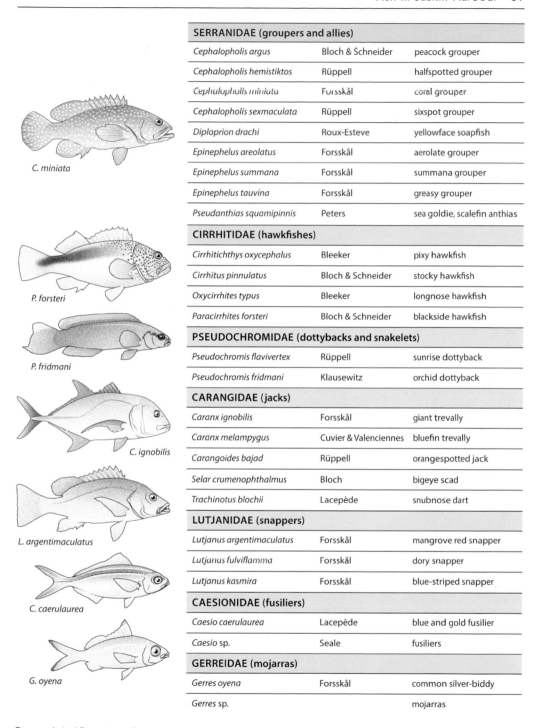

SERRANIDAE (groupers and allies)		
Cephalopholis argus	Bloch & Schneider	peacock grouper
Cephalopholis hemistiktos	Rüppell	halfspotted grouper
Cephalopholis miniata	Forsskål	coral grouper
Cephalopholis sexmaculata	Rüppell	sixspot grouper
Diploprion drachi	Roux-Esteve	yellowface soapfish
Epinephelus areolatus	Forsskål	aerolate grouper
Epinephelus summana	Forsskål	summana grouper
Epinephelus tauvina	Forsskål	greasy grouper
Pseudanthias squamipinnis	Peters	sea goldie, scalefin anthias
CIRRHITIDAE (hawkfishes)		
Cirrhitichthys oxycephalus	Bleeker	pixy hawkfish
Cirrhitus pinnulatus	Bloch & Schneider	stocky hawkfish
Oxycirrhites typus	Bleeker	longnose hawkfish
Paracirrhites forsteri	Bloch & Schneider	blackside hawkfish
PSEUDOCHROMIDAE (dottybacks and snakelets)		
Pseudochromis flavivertex	Rüppell	sunrise dottyback
Pseudochromis fridmani	Klausewitz	orchid dottyback
CARANGIDAE (jacks)		
Caranx ignobilis	Forsskål	giant trevally
Caranx melampygus	Cuvier & Valenciennes	bluefin trevally
Carangoides bajad	Rüppell	orangespotted jack
Selar crumenophthalmus	Bloch	bigeye scad
Trachinotus blochii	Lacepède	snubnose dart
LUTJANIDAE (snappers)		
Lutjanus argentimaculatus	Forsskål	mangrove red snapper
Lutjanus fulviflamma	Forsskål	dory snapper
Lutjanus kasmira	Forsskål	blue-striped snapper
CAESIONIDAE (fusiliers)		
Caesio caerulaurea	Lacepède	blue and gold fusilier
Caesio sp.	Seale	fusiliers
GERREIDAE (mojarras)		
Gerres oyena	Forsskål	common silver-biddy
Gerres sp.		mojarras

Figure 6.1 (Continued)

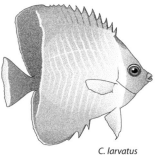

P. albovittatus

P. hamrur

L. mahsena

P. forsskåli

S. ghanam

P. oualensis

K. vaigiensis

C. larvatus

HAEMULIDAE (grunts and thick lips)		
Plectorhinchus albovittatus	Smith	giant sweetlips
Plectorhinchus gaterinus	Forsskål	blackspotted rubberlip
PRIACANTHIDAE (bigeyes)		
Priacanthus hamrur	Bleeker	goggle-eye
LETHRINIDAE (emperors)		
Lethrinus borbonicus	Valenciennes	snubnose emperor
Lethrinus harak	Forsskål	thumbprint emperor
Lethrinus mahsena	Forsskål	sky emperor
Lethrinus sp.		emperors
Monotaxis grandoculis	Forsskål	bigeye emperor
MULLIDAE (goatfishes)		
Mulloidichthys flavolineatus	Lacepède	yellowstripe goatfish
Parupeneus cyclostomus (prev. P. chrysendros)	Lacepède	gold-saddle goatfish
Parupeneus forsskåli	Fourmanoir & Guèzè	Red Sea goatfish
Parupeneus macronemus	Lacepède	long-barbel goatfish
Upeneus vittatus	Forsskål	yellowstriped goatfish
NEMIPTERIDAE (threadfin breams and whiptail breams)		
Scolopsis ghanam	Forsskål	Arabian monocle bream
PEMPHERIDAE (sweepers)		
Pempheris oualensis	Cuvier	blackspot sweeper
KYPHOSIDAE (sea chubs)		
Kyphosus vaigiensis	Quoy & Gaimard	brassy chub
CHAETODONTIDAE (butterflyfishes)		
Chaetodon auriga	Forsskål	threadfin butterflyfish
Chaetodon austriacus	Rüppell	exquisite butterflyfish
Chaetodon fasciatus	Forsskål	striped butterflyfish
Chaetodon larvatus	Rüppell	hooded butterflyfish
Chaetodon madagaskariensis	Ahl	Seychelles butterflyfish
Chaetodon melannotus	Bloch & Schneider	blackback butterflyfish
Chaetodon semilarvatus	Cuvier & Valenciennes	masked butterflyfish
Chaetodon trifascialis	Quoy & Gaimard	chevron butterflyfish
Heniochus intermedius	Steindachner	Red Sea bannerfish

Figure 6.1 (Continued)

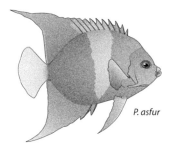

P. asfur

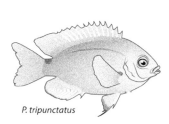

P. tripunctatus

POMACANTHIDAE (angelfishes)		
Centropyge multispinis	Playfair	multispine angelfish
Chaetodontoplus mesoleucus	Bloch	vermiculated angelfish
Pomacanthus asfur	Forsskål	Arabian angelfish
Pomacanthus imperator	Bloch	emperor angelfish
Pomacanthus maculosus	Forsskål	yellowbar angelfish
Pygoplites diacanthus	Boddaert	regal angelfish
POMACENTRIDAE (damselfishes)		
Abudefduf sexfasciatus	Lacepède	scissortail sergeant
Abudefduf sordidus	Forsskål	blackspot sergeant
Abudefduf vaigiensis	Quoy & Gaimard	Indo-Pacific sergeant
Amblyglyphidodon sp.		damselfishes
Amphiprion bicinctus	Rüppell	twoband anemonefish
Chromis caerulea	Cuvier	green chromis
Chromis sp.		chromis
Chromis ternatensis	Bleeker	ternate chromis
Dascyllus aruanus	Linnaeus	whitetail dascyllus
Dascyllus marginatus	Rüppell	threespot dascyllus
Dascyllus trimaculatus	Rüppell	threespot dascyllus
Neoglyphidodon melas	Cuvier	bowtie damselfish
Neopomacentrus sp.		lyretail damselfishes
Pomacentrus albicaudatus	Baschieri-Salvadori	whitefin damsel
Pomacentrus sulfureus	Klunzinger	sulphur damsel
Pomacentrus tripunctatus	Cuvier	threespot damsel
Pycnochromis dimidiatus	Klunzinger	chocolatedip chromis
Stegastes lacrymatus	Quoy & Gaimard	white spotted devil
Stegastes lividus	Bloch & Schneider	blunt snout gregory
Stegastes nigricans	Lacepède	dusky farmerfish
LABRIDAE (wrasses)		
Bodianus anthioides	Bennett	lyretail hogfish
Bodianus axillaris	Quoy & Gaimard	axilspot hogfish
Cheilinus fasciatus	Bloch	redbreasted wrasse
Cheilinus lunulatus	Forsskål	broomtail wrasse
Cheilinus trilobatus	Lacepède	tripletail wrasse
Cheilinus undulatus	Rüppell	humphead wrasse

Figure 6.1 (Continued)

LABRIDAE (wrasses, cont.)

Coris caudimaculata	Quoy & Gaimard	spottail coris
Coris cuvieri (prev. *Coris gaimard africans*)	Smith	African coris
Coris variegata	Rüppell	dapple coris
Epibulus insidiator	Pallas	sling-jaw wrasse
Gomphosus caeruleus	Lacepède	green birdmouth wrasse
Halichoeres hortulanus (prev. *Halichoeres centriquadrus*)	Lacepède	checkerboard wrasse
Halichoeres scapularis	Bennett	zigzag wrasse
Hemigymnus fasciatus	Bloch	barred thicklip
Labroides dimidiatus	Valenciennes	bluestreak cleaner wrasse
Larabicus quadrilineatus	Rüppell	fourline wrasse
Novaculichthys taeniourus	Lacepède	rockmover wrasse
Oxycheilinus celebicus	Bleeker	celebes wrasse
Pseudocheilinus hexataenia	Bleeker	sixline wrasse
Pteragogus pelycus	Randall	sideburn wrasse
Stethojulis albovittata	Bonaterre	bluelined wrasse
Thalassoma lunare	Linnaeus	moon wrasse
Thalassoma purpureum	Forsskål	surge wrasse
Thalassoma rueppellii	Klunziger	Klunzinger's wrasse

SCARIDAE (parrotfishes)

Cetoscarus bicolor	Rüppell	bicolour parrotfish
Chlorurus gibbus (prev. *Scarus gibbus*)	Rüppell	heavybeak parrotfish
Chlorurus sordidus (prev. *Scarus sordidus*)	Forsskål	daisy parrotfish
Hipposcarus harid	Forsskål	candelamoa parrotfish
Scarus ferrugineus	Forsskål	rusty parrotfish
Scarus ghobban	Forsskål	blue-barred parrotfish
Scarus niger	Forsskål	dusky parrotfish

BLENNIIDAE (blennies)

Aspidontus tractus	Fowler	mimic blenny
Ecsenius aroni	Springer	Aron's blenny
Ecsenius frontalis	Valenciennes	smooth-fin blenny
Ecsenius gravieri	Pellegrin	Red Sea mimic blenny
Istiblennius edentulus	Forster & Schneider	rippled rockskipper
Meiacanthus nigrolineatus	Smith-Vaniz	blackline fangblenny
Mimoblennius cirrosus	Smith-Vaniz & Springer	fringed blenny

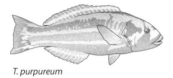
T. purpureum

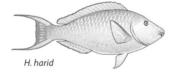
H. harid

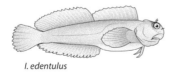
I. edentulus

Figure 6.1 (Continued)

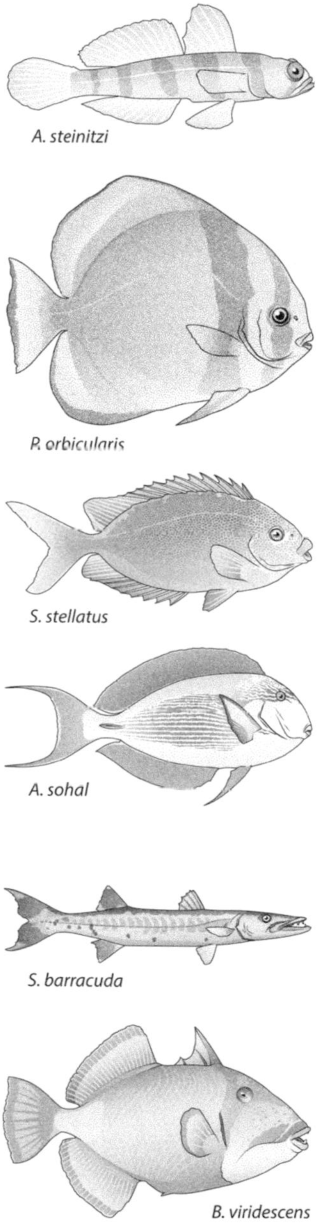

BLENNIIDAE (blennies, cont.)		
Petroscirtes sp.		combtooth blennies
Plagiotremus tapeinosoma	Bleeker	piano fangblenny
Plagiotremus townsendi	Regan	Townsend's fangblenny

GOBIIDAE (gobies)		
Acentrogobius sp.		gobies
Amblyeleotris steinitzi (prev. Cryptocentrus steinitzi)	Klausewitz	Steinitz' prawn-goby
Amblygobius albimaculatus	Rüppell	butterfly goby
Cryptocentrus caeruleopunctatus	Rüppell	Harlequin prawn-goby
Cryptocentrus lutheri	Klausewitz	Luther's prawn-goby
Ctenogobiops maculosus	Fourmanoir	Seychelles shrimpgoby
Koumansetta hectori	Smith	Hector's goby
Lotilia graciliosa	Klausewitz	whitecap goby

EPHIPPIDAE (spadefishes)		
Platax orbicularis	Forsskål	orbicular batfish

SIGANIDAE (rabbitfishes)		
Siganus stellatus	Forsskål	brown-spotted spinefoot

ACANTHURIDAE (surgeonfishes)		
Acanthurus gahhm	Forsskål	black surgeonfish
Acanthurus nigrofuscus	Forsskål	brown surgeonfish
Acanthurus sohal	Forsskål	Sohal surgeonfish
Ctenochaetus striatus	Quoy & Gaimard	striated surgeonfish
Naso sp.		unicornfishes
Naso unicornis	Forsskål	bluespine unicornfish
Zebrasoma desjardinii	Bennett	Indian sail-fin surgeonfish

SPHYRAENIDAE (barracudas)		
Sphyraena barracuda	Walbaum	great barracuda
Sphyraena jello	Cuvier	pickhandle barracuda

BALISTIDAE (triggerfishes)		
Balistapus undulatus	Mungo-Park	orange-lined triggerfish
Balistoides viridescens	Bloch & Schneider	titan triggerfish
Pseudobalistes flavimarginatus	Rüppell	yellowmargin triggerfish
Pseudobalistes fuscus	Bloch & Schneider	yellow-spotted triggerfish
Rhinecanthus assasi	Forsskål	Picasso triggerfish

A. steinitzi

P. orbicularis

S. stellatus

A. sohal

S. barracuda

B. viridescens

Figure 6.1 (Continued)

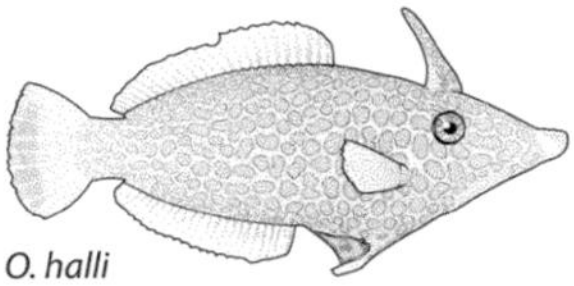

BALISTIDAE (triggerfishes, cont.)		
Sufflamen albicaudatum (prev. Sufflamen albicaudatus)	Rüppell	bluethroat triggerfish
Sufflamen chrysoptera (prev. Hemibalistes chrysoptera)	Bloch	halfmoon triggerfish

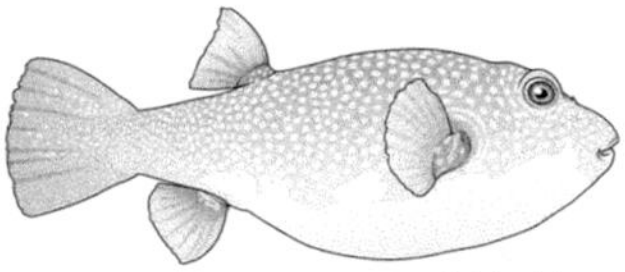

MONACANTHIDAE (filefishes)		
Oxymonacanthus halli	Marshall	Red Sea longnose filefish

OSTRACIIDAE (boxfishes and cowfishes)		
Ostracion cubicum (prev. Ostracion argus)	Rüppell	yellow boxfish
Ostracion cyanurus	Rüppell	bluetail trunkfish

TETRAODONTIDAE (puffers)		
Arothron diadematus	Rüppell	masked puffer
Arothron hispidus	Lacepède	white-spotted puffer
Arothron nigropunctatus	Bloch & Schneider	blackspotted puffer
Canthigaster margaritata	Rüppell	pearl toby

DIODONTIDAE (porcupinefishes)		
Diodon hystrix	Linnaeus	spot-fin porcupinefish

O. halli

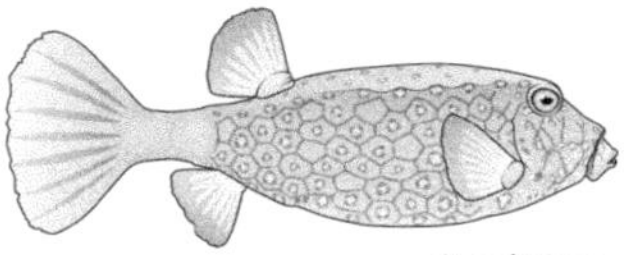

O. cubicum

A. hispidus

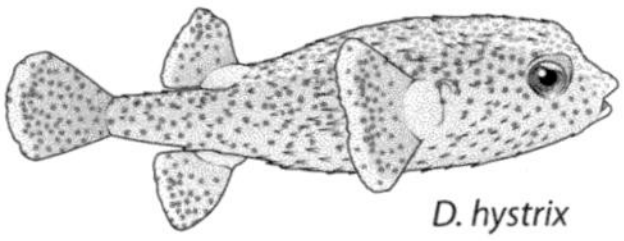

D. hystrix

Figure 6.1 (Continued)

Chapter 7

Fish in Dungonab Bay

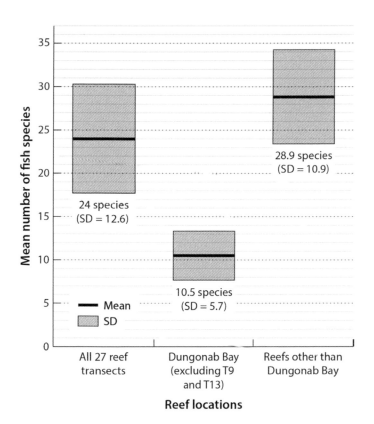

Figure 7.1 Diversity of fish species recorded in Dungonab Bay versus other reefs. The mean number of selected fish species recorded on the 27 reefs where fish were recorded is 24 [standard deviation (SD) = 12.6]. The mean number of selected fish species recorded in Dungonab Bay (excluding Transects 9 and 13) is 10.5 (SD = 5.7). The mean number of selected fish species on reefs other than those in Dungonab Bay is 28.9 (SD = 10.9).

DOI: 10.1201/9781003335795-8

Nine sites were studied in Dungonab Bay. Seven of these were reefs on which fish were recorded. Of these seven, the mean number of fish (from the list of selected species) was 10.5, which contrasts with the figure of 28.9 for other reefs. The bay has a greatly reduced hard coral fauna, and many coral-associated fish species were excluded. The most abundant coral genus is *Galaxea* (in contrast with its status on reefs outside the bay), and the most abundant chaetodon is *Chaetodon larvatus*, which ranks relatively low alongside other butterflyfishes from other reefs outside the bay. The main covering on rock surfaces is the soft coral *Xenia*, on which *C. larvatus* was seen feeding.

The planktivorous species *Pseudanthias squamipinnis* was also notable for its absence from the bay proper. This absence was presumably due to the absence of suitable cover among branching corals and to the relative paucity of plankton. On many dives within the bay, there appeared to be a preponderance of predatory species such as groupers (*Anyperodon leucogrammicus* and *Epinephelus merra* in particular) and barracudas (*Sphyraena barracuda*) and a paucity of smaller coral-associated species.

Chapter 8

Fish on Cousteau's Garage

Characteristically associated with the garage were a few fishes such as a large school of *P. squamipinnis* that was always present around the upper levels of the structure, concentrated on the side facing into the current. Mingling with these were several blennies (*Ecsenius midas*), whose similar red colour, crescentic tails, and prominent eyes indicated that it had evolved as a mimic of *Pseudanthias*. Planktivorous damselfish (*Chromis caerulea* and *Chromis ternatensis*) were also abundant. In terms of reef ecology, browsing and grazing species such as

Figure 8.1 Cousteau's Conshelf 2 project created an artificial reef with fresh settlement surfaces for corals and many other invertebrates. Many individual coral colonies were tagged and monitored from young specimens to large heavy and vulnerable *Acropora* tables. Fishes occupied similar niches to those of the natural reefs nearby.

© Frank Schneider/Alamy

DOI: 10.1201/9781003335795-9

Zebrasoma veliferum, Hipposcarus harid, Chlorurus sordidus, Thalassoma lunare, and *Parupeneus macronemus* were constantly feeding on organisms on the surface of the garage. Their feeding activities were so intense that the yellow paint that originally covered the garage was still quite visible in some areas in 1975.

During his underwater experiments, Cousteau employed his divers to keep the garage and house surfaces clean, and since then, fish have been grazing filamentous turf algae with similar effects. Vermeij et al. (2010) studied "The Effects of Nutrient Enrichment and Herbivore Abundance on the Ability of Turf Algae to Overgrow Coral in the Caribbean" and showed how turf algae is outcompeting live corals to the detriment of healthy reefs. Many young coral colonies were being removed by fish grazing on the surface of the garage, and probably only a small proportion of those colonies that settled became fully established. The dramatic effect that grazing fish have on reef ecology has been discussed previously by the author (Vine 1974).

Section II

Corals

Along with the diversity of fish comes a similar abundance of corals (Edwards & Head 1987). Scheer et al. (1983) produced a "Report on the stony corals from the Red Sea," which was published in *Zoologica*. It is largely a historical summary, ranging from T. Shaw's work in 1738 to the deep submersible dives by Fricke in 1981, when he collected corals from 200 m, including 161 species from 51 genera of reef-building corals and 33 species from 19 genera classified as ahermatypic. It is a significant piece of work, with descriptions of 194 species and illustrations of 176 species. Four species were described as new to science, and 22 species were found for the first time in the Red Sea. Veron JEN. has created a useful database at http://coralsoftheworld.org.

Berumen et al. (2013) review the status of coral reef ecology in the Red Sea and Arabian region (also see Eltayeb Ali et al. 2016; Head 1980), stating that the "Red Sea has the highest levels of endemism among all regions of the Indian Ocean . . . influenced by the unique environmental gradients of the Arabian region." They also state that the thermal bleaching events in the Red Sea highlight the pressures and challenges for future recovery. Coral diversity on Sudanese reefs is also discussed by Eltayeb et al. (2016) and El Hag (1999), who monitored corals in seven different areas extending from north to south, including Osef, Dungonab, Arikiyai, Arous, Abuhashish, Bashair, and the historic reefs of Suakin (see Greenlaw 1976). The results of transect studies, cluster analysis, and *t*-tests focused on 136 species of hermatypic corals belonging to 42 genera, with species richness ranging from 38 species in the Dungonab region to only 12.33 species in Bashair. The highest value for the index of similarity was 68 percent registered between Arikiyai and Suakin, whereas the lowest value was 32 percent registered between Suakin and Bashair.

DOI: 10.1201/9781003335795-10

Chapter 9

Coral Habitats
Building Sites and Graveyards

One of the most interesting long-term studies of coral reefs is that carried out by Tony Ayling (2023) at Snapper Reef in the northern Great Barrier Reef. Working in liaison with the Great Barrier Reef Marine Park Authority and the Australian Institute of Marine Science (AIMS), the Aylings monitored growth and recovery at marked locations around the island. They recorded a major event once every 3.5 years. From 1982 to 2022, they logged dramatic changes in coral cover at Snapper Island North Shallow. Coral cover was negatively affected by floods, cyclones, CoTS, bleaching, and algal growth. Impressively resilient, signs of recovery began quite soon after the corals were destroyed.

In 1982, they recorded only 10 percent coral cover, but this rose to 90 percent in 1993. Floods and bleaching followed by a cyclone caused the reef to almost disappear, again reaching only 10 percent cover in 2000. Recovery began again, attaining 60 percent cover in 2008, only to be hit again by cyclones, CoTS, and bleaching dropping to only 1 percent in 2015, and a new stage of recovery is now underway.

Fine et al. (2019) discuss the Red Sea's coral reefs in "Coral Reefs of the Red Sea – Challenges and Potential Solutions." Their study recognises the profound changes that are taking place north and south, along both eastern and westerly Red Sea coastlines. High temperatures and salinities are exacerbated by desalination plants, shipping, pollution, coastal towns, tourism projects, and acidification – the list of stressors is long and daunting. The environment, however, is not uniform. In planning the management of these threatened habitats, differences along the latitudinal gradient must be recognised (Loya 1976).

Reef communities fluctuate not only throughout the year but also over extended time frames during which cataclysmic events may cause extensive dieback of corals, algae, or grazing species such as *Diadema* sea urchins.

As the Ayling study so graphically illustrates, studies that reveal the processes of coral growth and decay necessitate data collection over several years. In the Red Sea, Reinicke et al. (2003) describe an opportunity to revisit Sanganeb 11 years (1991) after an initial survey (1980). These comparative surveys at Sanganeb have provided valuable data on the health and dynamics of Sudan's coral reefs (Klaus R 2015). In 1991, 42–60 percent of the total test areas, slightly more than in the 1980 census, comprised inanimate (unoccupied) substrata. Live actinian cover, mainly scleractinian and soft corals (alcyonarian) species, was analysed with regard to changes in community structures during the 11-year investigation period. The later survey indicated a relatively stable coral community, influenced, as in the 1980 survey, by the effects of different patterns of water movement as one moved from maximum exposure to most sheltered reef habitats. A total of 90 reef-building (scleractinian) coral species were identified during this Sanganeb-based

DOI: 10.1201/9781003335795-11

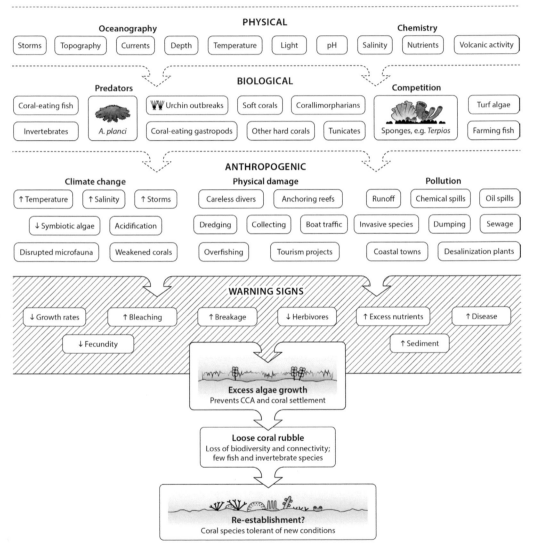

CORAL MORTALITY CASCADE

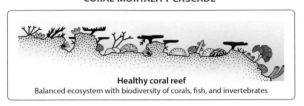

Healthy coral reef
Balanced ecosystem with biodiversity of corals, fish, and invertebrates

PHYSICAL

Oceanography						Chemistry			
Storms	Topography	Currents	Depth	Temperature	Light	pH	Salinity	Nutrients	Volcanic activity

BIOLOGICAL

Predators | Competition

| Coral-eating fish | | Urchin outbreaks | Soft corals | Corallimorpharians | | Turf algae |
| Invertebrates | A. planci | Coral-eating gastropods | Other hard corals | Tunicates | Sponges, e.g. Terpios | Farming fish |

ANTHROPOGENIC

Climate change | Physical damage | Pollution

↑ Temperature	↑ Salinity	↑ Storms	Careless divers	Anchoring reefs	Runoff	Chemical spills	Oil spills
↓ Symbiotic algae	Acidification		Dredging	Collecting	Boat traffic	Invasive species	Dumping	Sewage
Disrupted microfauna	Weakened corals		Overfishing	Tourism projects	Coastal towns	Desalinization plants

WARNING SIGNS

| ↓ Growth rates | ↑ Bleaching | ↑ Breakage | ↓ Herbivores | ↑ Excess nutrients | ↑ Disease |
| ↓ Fecundity | | | | ↑ Sediment |

Excess algae growth
Prevents CCA and coral settlement

Loose coral rubble
Loss of biodiversity and connectivity;
few fish and invertebrate species

Re-establishment?
Coral species tolerant of new conditions

Figure 9.1 Causes of coral mortality. Early physical, biological, and anthropogenic triggers can lead to a cascade of effects that ultimately degrade reefs, which shift to algal overgrowth and coralline rubble. Illustrations by Fiona Martin and Peter J. Vine.

© Vine

study. Identified zones included a *Xenia macrospiculata* zone, a *Lobophyllia corymbosa* zone, a *Sinularia-Dendronephthya* zone, and an *Acropora–Pocillopora verrucosa* zone.

Given the apparently different responses of corals in the mid- to northern Red Sea compared to corals on Australia's Great Barrier Reef, the question has to be asked, "what is it that renders them so distinct from each other?" An increase of only two degrees above summer temperature peaks is sufficient to trigger wide-scale bleaching on the Australian reefs, whereas corals from the Gulf of Aqaba can tolerate rises of up to seven degrees – a huge difference in physiological responses. Diurnal tides in the northern Red Sea have a considerably reduced vertical range compared to those off the GBR. Exposed low tides are therefore less of a threat to shallow corals. In addition, the floods that dump huge volumes of freshwater into coastal waters on the GBR do not occur in the northern Red Sea, and CoTS aggregations are relatively rare events in the northern Red Sea – much less consequential than anything seen on the GBR. (Kleinhaus et al. 2020).

Tony and Avril Ayling's invaluable year-by-year measurements on Snapper Reef, spanning 40 years, make a strong case for the importance of environmental factors playing vital roles in coral survival. The biggest dangers for reef recovery lie in loss of herbivores, mainly fish, and their vital role in controlling algal takeover of weakened corals. Could it be that northern Red Sea corals, stressed by temperature rise, are less challenged than their Australian cousins, heightening their resistance? Long-term surveys like that by the Aylings are helping to answer the question and offer hope for longer-term survival of coral reefs (https://youtu.be/e6JCQmN6_S8).

As our data from Shaab Rumi demonstrates, among the fastest-growing species to settle on the garage and colonise the new surfaces that it offered were table *Acropora* and *Porites* corals. Studies by AIMS scientists on Australia's Great Barrier Reef reached similar conclusions. Biologists studying reef regeneration showed that the presence of table corals can make a huge difference, with recovery rates up to 14 times higher when they are abundant. They point out that table forms have large, flat, plate-like shapes, providing vital protection for large fish in shallow reef areas and serving as shelters for small fishes, with some species almost entirely dependent on table corals. Reef management should take this into account, protecting the table corals from avoidable disturbance.

Chapter 10

Sea Urchins' Impacts on Coral Reefs

Diadema setosum is a black-spined sea urchin that plays a significant role in coral reef dynamics in the Red Sea (Dang et al. 2020) and elsewhere. Like a number of other urchins, they are important herbivores that graze on algae, often aggregating in densely packed communities that shade themselves from bright sunlight, by hiding during the daytime and emerging to graze on algae-coated rock faces at nighttime. In those habitats where its population is considered high (e.g., 16 individuals per m^2), the long black-spined urchins may gather in the open, feeding both day and night, scraping clean dead coral rock of its

Figure 10.1 Diadema setosum grazing on algal-coated coralline rubble. Removal of grazing sea urchins can lead to overgrowth by algae and death of corals.

© Vichai Phububphapan/Alamy

DOI: 10.1201/9781003335795-12

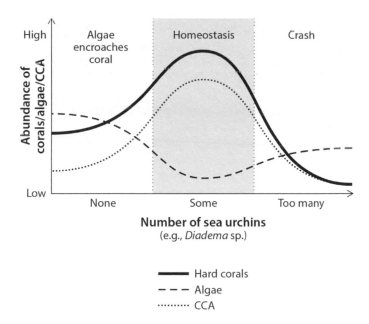

Figure 10.2 Urchins' Impact on Coral–Algal Dynamics. Sea urchins' impact on coral reefs can be quite dramatic, leading to a collapse of the essential balance of a healthy reef.

© Vine

algal turf coating, and, in the process, removing any coral planula larvae that settle. Where this occurs, it is possible to observe coral heads that are completely blanketed with *Diadema* urchins and dead coral heads that are scraped clean, bare of both macroalgae and young corals. This was observed during dives close to the entrance of Dungonab Bay and a short distance north of Port Sudan, where heads of *Galaxea* were frequently noted to be "covered in *Diadema*."

Population dynamics of the long-spined sea urchin, *Diadema savignyi*, have been studied in the Saudi Arabian Red Sea, where its significance as a keystone species in the regeneration of coral reefs has been clearly demonstrated. Unfortunately, their populations have been depleted in much of the Red Sea since there is a market for them as a food delicacy in parts of Asia. Efforts to cultivate nursery stock and to re-seed depleted reefs with young urchins have demonstrated the potential to repair reefs where reductions in herbivory have led to smothering algae such as Ulva. In addition to their ecological and economic importance, these urchins possess antimicrobial and anti-cancer properties, and this has focussed biologists' attention towards captive breeding, seed production, culture techniques, conservation strategies, and isolation of bioactive compounds. In the wild, the urchin's spawning peaks occur between September and October.

The presence, distribution, and impact of *Diadema* vary according to reef morphology and the presence of algal turf. Klaus commented on the "complete absence" of the common sea urchins (*Echinometra* spp. and *Diadema* spp.) from many reefs as a cause for concern. Previously, a Reef Check programme (Hodgson & Liebeler 2002) had noted that significant falls in the abundance of *Diadema* on coral reefs of the Indo-Pacific, including the Red Sea, between 1998 and 2001 corresponded with declines of reef-building corals (Carpenter et al. 2006). The absence of *Diadema setosum* from Sanganeb was repeatedly noted by (Reinicke et al. 2003).

Klaus did, however, note the abundance of *Diadema* in the northern Red Sea, "where it plays a major role in controlling reef development." Reef tops without urchins were, uncharacteristically for the Red Sea, overgrown with small macroalgae: a possible consequence of reduced grazing from herbivorous fish, such as parrotfish, and sea urchins.

Meanwhile, a study undertaken by Omri Bronstein and colleagues from Tel Aviv University's Steinhardt Museum of Natural History and School of Biology, released by the Royal Society in 2023, (https://doi.org/10.1098/rsos.230251) showed how *Diadema setosum* was an invasive species that had previously spread from the northern Red Sea into the Mediterranean where it initially flourished but then experienced a mass mortality. In 2023 the Gulf of Aqaba's previously healthy population of black-spined urchins suffered a similar population crash, infected with a ciliate parasite like that recorded in the Caribbean and elsewhere in Arabian waters. This time the entire population of urchins was reported to be annihilated, presaging potentially disastrous consequences for the very reefs that biologists had considered a vital refuge for heat resilient corals.

The impact of *Diadema* on coral reefs is difficult to overestimate. A study in Taiwan (Dang et al. 2020) investigated the impact of sea urchins on algal abundance and coral recruitment. Researchers set up an experiment in which different densities of the urchins were entrapped by plastic netting cages attached to the reef. The trials confirmed the strong influence that *Diadema* may exert on algal and/or coral cover. They demonstrated that, while algal grazing keeps the coral rock "clean" and suitable for settlement of coral larvae, it also removes coral larvae, suggesting that any efforts to control weed growth by introduction of sea urchins should take care not to overload the reef with urchins, as such overload would prevent new corals from settling. A Caribbean study published in 2001 by Peter Edmunds and Robert Carpenter indicates that *Diadema* may not always be a negative factor in coral recruitment and the re-establishment of coral-dominated reefs. These researchers point to a stretch of coastline in Jamaica where a *Diadema*-grazed zone, 60 m wide, lacked macroalgae, but where the density of *Diadema antillarum* was 10 times higher, and the density of juvenile corals up to 11 times higher than in the seaward algal zone. They suggested that the *Diadema* played a positive role in coral settlement and reef recovery (Edmunds & Carpenter 2001). *Porites* and *Agaricia* corals tended to be associated with recovered populations of *Diadema*, while *Acropora* and *Montastraea* seemed less able to re-establish themselves in the *Diadema* zone.

Chapter 11

Coral Communities

Corals are made of many different elements, and the sum of all these parts, referred to as the coral holobiont, influences how the coral's life systems operate (Sharp et al. 2012), including food webs, life cycles, and chemical and nutrient cycling. Simply put, corals are multidimensional individuals – meta organisms – working in tandem to create highly efficient carbon factories. We are familiar with symbiotic algae (zooxanthellae) that live within the coral cells, but perhaps less aware of the critical role played by microbes. Microbes regulate cells' health and productivity, helping to ensure larval recruitment, bacterial colonisation, and the presence or absence of diseases – in short, they are the control centres for the coral biont. We need to look

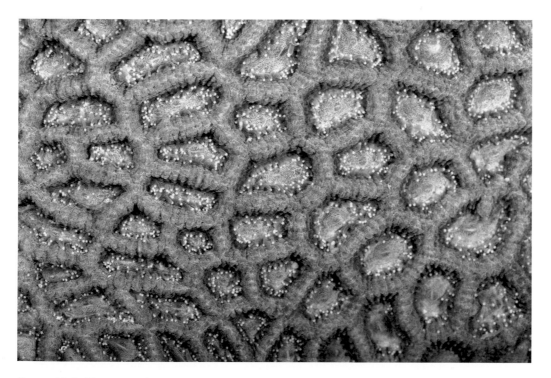

Figure 11.1 Close-up of coral polyps feeding at night.
© Hans Sjöholm

DOI: 10.1201/9781003335795-13

here when analysing a species' response to changing physical factors such as rising temperatures and acidification of seawater – both of which may be attributed to climate change. How well coral species survive widespread alterations to their habitats will depend, in turn, on how the holobiont reacts. Do its various microbionts wither and die, or can they tweak their controls, adapting into beings that are able to function at higher temperatures or greater acidities and thus contributing to both the species' recovery from attack and long-term survival?

When we observe the broad picture of coral settlement, growth, and survival in the Red Sea, our studies should drill down (Kötter 2001), through these layers of complexity, to decipher the overall ecology at all levels. Few studies have done this. Mergner and Schuhmacher (1985) built on earlier studies aimed at quantifying coral communities at Sanganeb atoll. Berumen et al. (2019) review the status of our knowledge of Red Sea corals, and DeVantier et al. (2000a) study coral communities of the central-northern Red Sea in Saudi Arabian waters. Sampling at 10 m depth on the windward and leeward sides of Sanganeb reef, Klaus (2015) and Schuhmacher's studies neatly demonstrate the powerful influence of water movement in coral settlement, growth, and survival. They propose a number of zones that could be defined in terms of water movement, such as the outer northeast side of the reef ring, which was exposed to strong swell and surf; the inner side of the south ring, which was met by a strong long-reef current; the outer lee of the southwest atoll ring; and the inner lee of the northeast side.

As conditions change, so do the reefs themselves. Cacciapaglia and Van Woesik (2015) sought to identify reef coral refugia that might play roles in coral survival as the oceans go through the dramatic changes associated with climate change, including thermal stress on reef corals in the Red Sea. Of the 12 coral species selected for their abilities to adapt, 24–50 percent of their current habitat was predicted to be lost by 2100, primarily between the latitudes 5° and 15°, in both hemispheres. However, when these researchers factored in a 1°C capacity to adapt, "two ubiquitous species, *Acropora hyacinthus* and *Acropora digitifera*, were predicted to retain much of their current habitat." In all, five areas were identified as Indian Ocean refugia, all of which are highlighted for their importance in conservation management (Sheppard 2012).

Chapter 12

Coral-Killing Sponges

Cousteau (1971) described the situation at the island of Europa in the Indian Ocean in his typically graphic manner as the reef being covered "with a greenish glue-like substance" and "a stretch of puny, grayish coral, some of which was covered with that green mucus" and during one of those dives Chauvellin, Jaubert, Deloire and Coll were filming turtles when they discovered an area that they immediately named the "demolition area." It was a relatively large field of dead coral. Even worse, there seemed to be little, if any, fauna of any kind. Not a sea fan nor a fish . . . it was a field of grayish dead coral.

The biological horizon was changing before my eyes. While the general public had become aware of threats from coral-eating starfish, other veiled threats were going unnoticed. Studies on the spread of coral-killing sponges, such as the enigmatic *Terpios* cf. *hoshinota* (Shi et al. 2012; Montano et al. 2015), failed to trigger warning alarms. Even today, this coral killer, lacking the notoriety of CoTS, has hardly attracted public attention, despite presenting significant danger to coral reefs across a broad region of the tropics (Das 2020).

My studies on *T.* cf. *hoshinota* demonstrated the importance of longer-term recording throughout the year (Vine & Head 1977). Some affected corals in winter were seen to "fight back" against the potentially smothering sponge (Syue et al. 2021). The driving force in these interactions is a battle for space, with sponges tending to gain the upper hand in summer, while the opposite occurs under lower sea temperatures in winter (Nozawa et al. 2016). In the short to medium term, this phenomenon generally results in the death of corals over large sections of the reef, creating a knock-on effect that results in the proliferation of macroalgae and the disappearance of many reef fishes (Diaz-Pulido et al. 2010). There is evidence that, in the longer term, CCA can suppress macroalgae and sustain footholds for new settlement and growth by more resilient corals, forcing a retreat by the invasive sponges.

Cousteau described several encounters with what appears to have been *Terpios* in 1967–1968 in the Red Sea (Cousteau 1971). Bryan (1973) was one of the first scientists to report on the sponge–coral interface. The presence of the sponge in the garage (where it was observed to kill an *Acropora* table and some other corals) was noted during research in 1972 and was reported as *Terpios* sp. in Vine and Vine (1980). Following a series of studies describing how the sponge was instrumental in triggering the decline of corals and their replacement by macroalgae, Reimer et al. (2010) provided a degree of hope. They noted that coral communities impacted by *T. hoshinota* could, if environmental conditions were right, return to coral dominance.

Hope also came with an understanding of CCA's role in coral ecology. Its role in cementing dead fragments was described by Vine (1974), and since then, the vital role of herbivory on

DOI: 10.1201/9781003335795-14

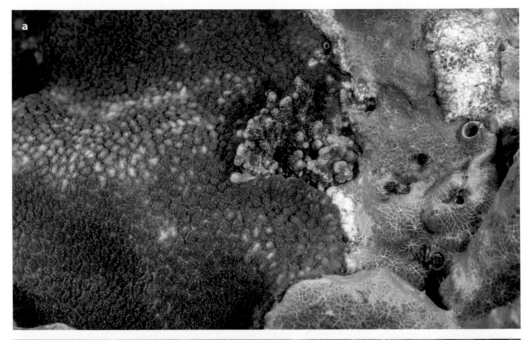

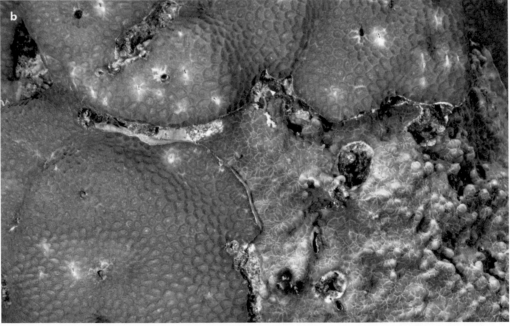

Figure 12.1 (*a, b*) Three protagonists in a contest for survival on the reef. In both panels *a* and *b*, hard coral (cf. *Echinopora* sp.) is on the left, pinkish CCA in the middle, and *Terpios* cf. *hoshinota* on the right. The coral is a prime reef builder, and CCA helps counteract sponge growth, encouraging coral larvae to settle, while the *Terpios* is a destroyer of the reef.

© Vine

coral reefs has been inextricably linked to CCA. Harrington et al. (2004) drew attention to its vital role in facilitating coral larval settlement. They also drew attention to the resilience of CCA when confronted by *T. hoshinota* commenting that:

> We have rarely seen *T. hoshinota* overgrow CCA, and contact between the two has been reported to cause retrogression in the sponge (Plucer-Rosario 1987). Thus, the availability of CCA in communities highly impacted by *T. hoshinota* could play an even more critical role in controlling the spread of the sponge by supporting stony coral recruitment and survival.

Hope notwithstanding, *Terpios* was spreading, seeming to respond to an invisible switch that, metaphorically speaking, turned brightness into darkness – colourful carbon-gobbling corals into brown macroalgae. Madduppa et al. (2017) and Elliott et al. (2016) investigated the spread of *T. hoshinota* and its role in the degradation of the reefs. Expressing a sense of alarm felt by the growing number of scientists who encountered widespread destruction by this species, they were emphatic in their introduction to their findings: "*Terpios hoshinota* is an encrusting sponge and a fierce space competitor. It kills stony corals by overgrowing them and can impact reefs on the square kilometre scale."

They also documented the retreat and advance of corals relative to sponges.

As the present author reported in his initial statements on the impact of *Terpios* (Vine & Vine 1980), the medium-term outcome of a conflict between hard coral and *T. hoshinota* may be challenging to predict. Aerts (Aerts 2000; Aerts et al. 1997) reported on the dynamics, stating there may be no clear winner, with both sponges and stony corals alternately losing and gaining space when observations were done over months. Successful fightback against sponge attacks is reported by Aerts and Van Soest (1997), McKenna et al. (1997), Schönberg et al. (2001), and Rützler (2004). Reports of this sponge overgrowing corals, starving them of light, and poisoning their tissues have become more frequent in recent years. Documentation of large areas of the reef being devastated on a scale comparable to that of *Acanthaster* outbreaks has

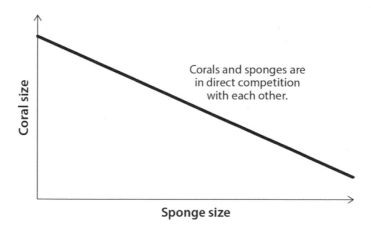

Figure 12.2 Coral–sponge dynamics.
© Vine

provided evidence of the impact of this coral killer (Haywood 2019) not just in the Red Sea but across the Indo-Pacific Ocean (Bryan 1973; Vine & Vine 1980; Plucer-Rosario 1987; Rützler & Muzik 1993; Fujii et al. 2011; Shi et al. 2012; de Voogd et al. 2013; Hoeksema et al. 2014; van der Ent 2016). As the present author observed in the Sudanese Red Sea, *T. hoshinota* can turn square kilometres of healthy coral reef, where sponges had been kept under control by a range of predators, into what looks like the remains of a massive disease outbreak. Syue Siang-Tai addressed the question of why *Terpios* kills hard corals, describing a series of field experiments, and concluded that "our results support the hypothesis that the encrusting sponge did not kill corals for food or nutrients, but rather for the substrate" (Newman & Dana 1972; Syue et al. 2021).

As with the CoTS story (Vine 1970, 1973), in which humans are frequently blamed as the root cause of these events, *T. hoshinota* outbreaks have been linked to various forms of pollution (Rützler & Muzik 1993) and iron enrichment of seawater. It should be noted that *Terpios* is by no means the only genus of sponge destroying coral reefs by smothering live corals.

The observations recorded in Cousteau's book *Life and Death in a Coral Sea*, about voyages made in 1967–1968, provide some of the earliest records of the coral-killing sponge *Terpios hoshinota* in the Red Sea and Indian Ocean. Meanwhile, notes on my personal encounters with *T.* cf. *hoshinota* in the Red Sea throw some light on the known history of its spread on Red Sea reefs. During my time in Sudan, quite suddenly, in 1973, I realised that something was going wrong with the country's coral reefs. Substantial sections of reefs were dying before my eyes, smothered by the inexorable extension of deadly thin layers of a dark grey sponge. Despite Cousteau's quite graphic descriptions, I did not initially make the connection to the killer sponge that had captured my attention, and I turned to the scientific community for help. Urgent enquiries were mailed to various scientific colleagues working in the field of coral reef ecology. The abiding questions were: did the sponge have a name, and if so, what did we know about it?

On 1 October 1975, I wrote to the Curator of Sponges at the British Museum of Natural History, but times were hard, and the response was discouraging: "Due to lack of staff, we have had to restrict our identification service to the British fauna only." Undeterred, we sent out an urgent appeal to the tropical marine biology diaspora, asking for advice. It was soon apparent that, while we were not the first to report on sponge-induced mass coral mortality, like other biologists, we were finding it hard to identify the culprit sponge or find detailed descriptions of its biology and ecology. In response to my letter to Prof. Joseph H. Connell from the Department of Biological Sciences at the University of California, Berkeley (21 October 1975), he confirmed the situation in a letter (23 January 1976) in which he wrote, "I can't think of any references to sponges smothering corals," but promised to discuss with his colleagues and counterparts. I also received responses from Prof. Willard D. Hartman at the Peabody Museum of Natural History, Yale University (21 January 1976), and from Henry Reiswig at the Redpath Museum, McGill University, in Montreal. Hartman wrote of the "blackish encrusting sponge that we both observed in Guam overgrowing living corals." It sounded as if Hartman's Guam sponge was at least closely related to our Red Sea species.

His notes read:

August 9 '71 Anae Is., first trip:

Purple-grey killer sponge all over in patches up to 10 metres across; is killing and covering living coral at the front of contact; direct contact without a clear dead area between sponge and coral.

You could have counted the number of sponge experts on one or two hands in those days, and they kept in close touch with each other. Thus, a letter from Gerald Bakus at the University of Southern California in December 1975 only reinforced the impression that very little was known about the coral-killing sponge that Cousteau had described in his popular book, and which had been tentatively identified as a species of *Terpios*. "I can recall only one instance where I saw a coral that was covered with a sponge. It was a large sponge . . . about 6–10 ft in diameter . . . located in the pass at Fanning Island."

Notwithstanding the paucity of taxonomic data, it was apparent that by the early 1970s Sudan's reefs were showing signs of stress, often manifested by extensive growths of the encrusting sponge and mass mortality of corals. A survey of coral growth on Cousteau's garage at Shaab Rumi provided a valuable window into what was happening on the reefs themselves.

Between visits made to Shaab Rumi in April 1975 and February 1976, many corals in the garage died or suffered damage. Of the bracket and table-form *Acropora* specimens, 44 percent suffered some form of injury, and the death toll was 25 percent of the total. Where possible, the cause of death was recorded, but this was not possible in some cases. By February 1976, all the colonies noted as being in poor condition or partially dead during April 1975 had completely died and, in some cases, broken off from the garage.

What was the cause of these large-scale die-offs? Coastal reefs close to urban developments, such as Port Sudan and Jeddah, were likely to be impacted, to a greater or lesser extent, by chemicals, oil, sewage (Reopanichkul et al. 2009; Wear & Thurber 2015), and waste dumping from ships. Still, it was not clear that any of these factors could explain the levels of mortality that we witnessed. The most significant cause of coral mortality was more likely to be a rise in sea temperatures, which caused a weakness in corals and made them susceptible to attack by species like *Terpios*, which was rapidly gaining an unsavoury reputation as a coral killer. However, predicting coral responses to varied stressors is not a straightforward science (Harborne et al. 2017).

I wrote about the coral-killing sponge (*Terpios* sp.) in my general paper on the ecology of Sudanese reefs (Vine & Vine 1980) and in an article on corals in Cousteau's garage (Vine & Head 1977), but the publications had limited circulation, and it is not surprising that many researchers missed them. By the time the killer sponge was finally described and named, field observations of its spread around the globe had multiplied to levels that demanded scientific attention.

My correspondent from the 1970s, Klaus Rützler, together with his colleague, Katherine Muzic, finally published a taxonomic description of the new species, *Terpios hoshinota*, in 1993 (Rützler & Muzic 1993). In doing so, they drew attention to a general increase, since the late 1960s or early 1970s, in reports of destructive agents affecting the world's coral reefs, including a number of causes, both natural and anthropogenic. One of the most impressive of such pests, at least in terms of its killing capacity, was their new species from the genus *Terpios*, which they named *Terpios hoshinota*. Many records were new ones based on the first underwater records by SCUBA divers seeing (or recognising) the sponge for the first time (Liao et al. 2007). The initial record, from Guam (Bryan 1973), was followed by many records, right across the Western Pacific in subtropical and tropical locations. Sometimes characterised as the "black disease," it has recently emerged as a significant threat to Indonesian reefs (de Voogd et al. 2013; van der Ent 2016) and is widely regarded as a species of interest throughout its range (Aini 2021).

Notwithstanding an ongoing debate concerning taxonomy and nomenclature, there is less ambivalence concerning the damaging impact of the dark grey encrusting thin sponge on coral

Figure 12.3 (a) *Leptoria phrygia* being smothered by *Terpios* cf. *hoshinota*. (b) *Favia speciosa* being attacked by *Terpios* cf. *hoshinota*.

© Vine

reefs, especially those already under stress (Rützler 2004). It frequently delivers the *coup de grâce* on reefs that are suffering from environmental degradation, be that of a general ecological nature such as an increase in SST or acidification (Miller et al. 2013) – both aspects of global warming – or of a more direct anthropogenic nature such as pollution, overfishing, or physical disturbance from anchoring (Yara 2011).

T. hoshinota was not the only coral-killing sponge that we observed in the Sudanese Red Sea. There were at least four species actively involved in this process, growing on Cousteau's garage: an unidentified green species (possibly *Dysidea* sp.), a grey encrusting species (*Terpios hoshinota*), an orange encrusting species (*Spirastrella* cf. sp.), and a red boring sponge (*P. vastifica*). On sides C to D of the garage, above and around the large table of *A. pharaonis* (colony code number 40), it was possible to see the lateral extension of *Terpios* during the 10-month period between visits. This encompassed a number of coral colonies that had been alive on my previous visit but were dead by February 1976 (Vine & Head 1977; Syue et al. 2021).

Certain colonies on the garage showed notable resistance to attack and remained as isolated living coral colonies surrounded by the encrusting grey *Terpios* sponge. It was considered likely that further extension of the sponges (particularly *Terpios* and *Spirastrella*) would occur and cause further mortality of corals on the garage. The phenomenon of corals being killed by

Figure 12.4 A red boring sponge, *Pione* cf. *vastifica*, hosts a range of species in its water channels and creates new surfaces on the reef, enabling the settlement of species that might otherwise be excluded by dense colonisation of corals and other species (Risk & Muller 1983). Many sponges live within the cavities of live corals and have been studied using endoscope technology.

© Vine

Figure 12.5 Sponges, soft corals, and hard corals grow on a cable at Chicken Wreck, formerly known as *M.V Glaros*.

© Hans Sjöholm

sponges was at that time an important but generally underrated factor in coral ecology. Previous investigations by the present author demonstrated that coral-killing sponges tend to grow rapidly in the summer months, when they may smother some Red Sea corals, but that sponge growth slows down considerably in the winter, and at this time, the sponges may regress and partially dead coral colonies may recover (Bell et al. 2013).

Chapter 13

Other Coral Killers

Unfortunately, the future of corals is quite bleak, thanks to global warming. Coral diseases in the Red Sea, especially prevalent in the summer months, have been reviewed by Mohamed and Sweet (2019) and Hazraty-Kari et al. (2021). Up to 40 coral diseases affecting Red Sea corals, such as black band disease, skeletal eroding band, pink line syndrome, and ulcerative white spot, are recognised. Coral mortality may be triggered by a variety of stress factors, including climate change-related disturbances (e.g., more frequent and more powerful hurricane-force winds, raised temperatures, pH shifts) and other natural causes (Birrell et al. 2008), such as volcanic ash, the aftermath of volcanic activity, and damage by divers (Abou Zaid & Kotb 2000).

Figure 13.1 *Acanthaster planci*, the crown-of-thorns starfish, feeding on live corals in the Sudanese Red Sea. The starfish is responsible for the deaths of many reefs, leaving rubble in its wake, but is not the only mass killer on the reefs.

© Vine

DOI: 10.1201/9781003335795-15

Aronson et al. (2003) discussed infectious diseases in corals and emphasised the limited state of our knowledge at that time. Indeed, according to their study, the first report of coral disease was published as recently as 1973 by A. Antonius, since when "only four coral diseases have been characterized in terms of microbial pathogens and, to a lesser extent, pathogenesis: black band disease, white band disease type II, plague type II, and aspergillosis." Despite the brevity of the scientific record, there is nothing particularly new about coral diseases themselves. In at least one case, a disease outbreak was geologically dated to at least 3,000 years ago. There is however an acceleration and spread of coral diseases that may be linked to a range of criteria, including, among others, rises in sea temperatures, pollution and coastal development.

Charles Sheppard (2016) posed the question: "Coral Reefs in the Gulf Are Mostly Dead Now, But Can We Do Anything About It?" He argues that we must make stronger efforts to control the processes that turn coral reefs into algal reefs. Observations on Sudanese reefs, reported in this chapter, record coral killing by certain sponges, aggregations of coral-eating invertebrates such as CoTS (*Acanthaster. planci*), and a rise in SSTs and salinities.

Such attacks on live coral colonies frequently result in the exposure of dead skeletal material, providing new settlement surfaces and giving rise to the spread of algae such as *Turbinaria*, accompanied by a reduction in the local biodiversity of fish and invertebrates. A study on Australia's Great Barrier Reef (Birrell et al. 2008) concluded that "the ecological resilience of coral reefs depends critically on the capacity of coral populations to re-establish in habitats dominated by macroalgae."

Figure 13.2 Crinoid on dead coral.
© Hans Sjöholm

The phase shift from coral domination to macroalgal domination may be more nuanced than I have indicated, with a variety of species capable of overgrowing reef building corals (Lang-mead & Chadwick-Furman 1999). For example, the corallimorpharian *Rhodactis rhodostoma* forms aggregations that dominate patches on some coral reef flats in the Red Sea and may represent a stage towards the shift in dominance from coral to macroalgae. Its polyps can over-grow zoanthids, hydrozoan corals, sponges, and encrusting macroalgae. Reef builders such as *Porites*, *Acropora*, and *Pocillopora* have also been observed to suffer from stinging nematocysts on the sweeping tentacles of this corallimorpharian.

Not all species are vulnerable to *Rhodactis*' stinging shield, however. Corals in the families Faviidae and Mussidae seem to be able to resist this corallimorpharian's sweeping tentacles, which elicit sharp responses when coming closer than 1–3 cm to these corals.

Chapter 14

Coral Bleaching

The year 1998 was "*annus horribilis*" for the world's coral reefs, with some areas suffering as much as 90 percent mortality due to bleaching. This devastation was spread right across the Caribbean and Indo-Pacific, with notable damage to reefs along Australia's Great Barrier Reef. The initial view was that affected reefs would disappear by 2050, but subsequent research revealed a growing pattern of corals surviving extreme summer temperature peaks in the northern Red Sea and the possibility that certain corals in such areas were developing, or had already developed,

Figure 14.1 Coral bleaching on a branching coral, *Acropora* sp.
© Steph Skermer/Stockimo/Alamy

DOI: 10.1201/9781003335795-16

resistant strains capable of tolerating temperature shocks where comparable rises elsewhere would be expected to lead to catastrophic bleaching (Furby 2013).

Scientists from the Institute of Earth Sciences at the University of Lausanne, studying corals from the Gulf of Aqaba, reported that a wide range of coral species growing there had tolerated rises of up to 7°C in sea surface summer temperatures, while a jump of only 2–3°C was enough to trigger bleaching in corals on Australia's Great Barrier Reef. The discovery has been attributed to the geological formation of the Red Sea, which was cut off from the Indian Ocean as recently as eight or nine thousand years ago, towards the end of the last Ice Age. As an enclosed body of saline water, only the most heat- or salt-tolerant species would have survived and thus evolved into the resilient strains that we see there today. It sounds hopeful in terms of the northern Red Sea acting as a refuge that could help to save our reefs in the future.

The aftermath of coral bleaching, like that of mass CoTS aggregations, generally involved dramatic transformations such as surges in fleshy algae that smothered the reef's healthy corals. In such cases, the colourful reefs became monochromatic – turning initially bright white but later coated in green algae, crustose coralline algae, sponges, and soft corals. Diversity diminished as a whole, including the typical list of fishes that were normally such prominent aspects of reef life.

Figure 14.2 Coral reefs have always been in a state of change: changing species, changing communities, changing levels of stress, both natural and anthropogenic, and also changing scales on local, regional, and global levels. The devastating phenomenon of coral bleaching falls into this bio narrative, whose theme is dominated by death and destruction on the reefs.

© Hans Sjöholm

Figure 14.3 Seen from above, thanks to the dazzling whiteness of corals that have lost their symbiotic intracellular algae, bleaching exposes the calcareous intricacies of the formerly vibrant coral forms.

© Hans Sjöholm

Figure 14.4 I recall diving off Ras Muhammed on a filming recce planned to identify flourishing reefs when the overwhelming impression was of an undersea graveyard with piles of dead coral rubble piled up into grim mementoes of what had until recently been regarded as some of the most impressive reefs in the world.

© Hans Sjöholm

Figure 14.5 When corals are stressed by changes in conditions such as temperature, light, or nutrients, they expel the symbiotic algae living in their tissues, causing them to turn completely white. Even after they have suffered bleaching, they are not necessarily dead but are more likely to succumb to the experience.

© Hans Sjöholm

The faster the change in temperature, the greater the impact on living corals, but there were other causal combinations of factors such as eutrophication, acidification, depth variation, and distance from shore.

In addition to the increased resilience of heat-resistant corals, some corals are more suscepti-ble to bleaching than others, and the impact of bleaching may lead, subject to rates of tempera-ture rise or degrees of pollution, to changes in coral assemblages rather than the outright death of all species.

Bleaching that occurs in reef-building corals, generally in response to a rise in local SST of 1–2°C above normal (6°C in the northern Red Sea) or unusually intense solar radiation (Jones & Hoegh-Guldberg 2001), is accompanied by a breakdown in the symbiosis between corals and their unicellular dinoflagellate symbionts (Weis 2008). Weis studied the processes that take place when corals lose their symbionts and expose their white skeletons. She characterises bleaching as a "host innate immune response to a compromised symbiont, much like innate immune responses in other host–microbe interactions."

Eakin et al. (2019) reviewed the 2014–2017 global-scale coral bleaching event, summaris-ing insights and impacts. Coral bleaching in the Sudanese Red Sea was not observed during the period of the author's reef surveys (Vine & Vine 1980) in 1973–1974 but has become more prominent since then. It has been established that nutrient-supplying ocean currents tend to exacerbate susceptibility to bleaching, suggesting that weather prediction models could be adapted to forecast bleaching.

While bleaching was not noted as a dominant factor affecting Sudan's reefs in the 1970s, cor-als on fringing and offshore reefs in the Dungonab area were observed to bleach in 1998, and the reefs were reported to be "so bright white as to be visible from the shore," while the weather conditions were "calm and very hot." The first systematic surveys in Sudan following the 1998 event were carried out along a 70-km section of coast in the Dungonab area to the north of Port Sudan in 2002 as part of the baseline surveys for the establishment of Dungonab Bay–Mukka-war Island Marine National Park (Salam 2006).

By comparison, many reefs south of Port Sudan and offshore, showing little evidence of widespread impacts, still supported a high cover of flourishing hard corals, including healthy communities of what are generally regarded as bleaching-susceptible genera, suggesting that these reefs may have been from the temperature-resistant strain that we have discussed earlier. Patterns in coral cover were reflected in the abundance of coral-eating butterflyfishes and other groups. Schuhmacher et al. (2005) studied the aftermath of bleaching as it occurred on a reef in the Maldives, including coral larvae settlement and reef regeneration.

The widespread coral mortality observed on the northern reefs was concentrated at depths of less than 10 m, in contrast to coral mortality caused by CoTS outbreaks in, for example, Sinai, where outbreaks often begin at depths as great as 50 m or more and cause extensive coral mor-tality at similar depths.

Coral bleaching is not a new phenomenon. Kamenos and Hennige (2020) comment on their reconstructed model of four centuries of temperature-induced bleaching on the Barrier Reef. Key questions in relation to coral bleaching are whether corals can resist bleaching and whether they can recover after a bleaching event. Much interesting work at the cellular and genetic lev-els has been conducted in recent years. Encouraged by the survival of individual coral colonies during mass coral bleaching events, Barshis et al. (2013) suggest that "some groups of cor-als may possess inherent physiological tolerance to environmental stress." They point out that some high-temperature environments naturally retain healthy, growing coral populations and that these corals can show elevated bleaching tolerance and thus be most likely to cope with future climate change. As such, they may offer essential information regarding observed differences in coral physiological resilience (Haywood 2019).

Decay of corals due to bleaching depends on the frequency and severity of mass bleaching episodes; the greater the bleaching episodes, the more likely it is for corals to be affected. *Porites*, *Seriatopora*, *Stylophora*, and *Acropora* are not only some of the fastest to exhibit bleaching but also among the quickest to show signs of recovery.

There are two kinds of coral bleaching: one associated with all-too-familiar mass coral bleaching events and the other associated with seasonal loss of algae and/or pigments from corals in deeper water. The latter is of particular interest and has been described for the mesophotic zone, from 40 to 63 m depth, of the northern Red Sea, where colonies of *Stylophora pistillata* experienced bleaching during the warmer months of April to September but began to recover their symbiotic algae and colour in October (Nir et al. 2014).

In experiments involving transplantation of *Stylophora* colonies, Nir and colleagues were unable to attribute a single environmental factor (e.g., temperature, light intensity, or food availability) to responsibility for seasonal bleaching. It was apparent, however, that mesophotic zooxanthellae are more successful at meeting metabolic requirements of the reef-building corals when chlorophyll *a* concentration decreases by over 60 percent during summer and early autumn (Nir et al. 2014).

Chapter 15

Growth and Distribution of Corals on Cousteau's Underwater Garage

Cmd Cousteau described Conshelf 2 as a "luxury operation," at least for the divers who spent a month at Starfish House, ten metres below the surface of the Red Sea. Several pieces of underwater equipment were left *in situ* on the reef, and an underwater garage was prominent on the reef terrace, just as it had been in 1963.

This fascinating underwater living experiment (Crylen 2018) provided an unexpected addition to the mass of scientific information that was collected during the month when French divers occupied its underwater habitats. Returning in 1968 to the site of the epic experiment, Cousteau (1971) remarked on the site's legacy for research studies on coral growth rates. All these structures were colonised by marine life (Figure 15.1). Cousteau noted that one small *Acropora* table coral growing on the garage now measured over 20 cm across, suggesting a faster growth rate (Ting-Ying 1934, 1959) than most had expected, but was in line with coral growth studies described in Buddemeier and Kinzie (1976), who reviewed the topic, and our own observations reported in the following.

Figure 15.1 The futuristic Conshelf 2 garage at Shaab Rumi has become a living laboratory for studying corals and other settlement species.

© Vine

DOI: 10.1201/9781003335795-17

Many of the flourishing corals on the submarine garage appeared to have initially settled in protected situations among crevices or irregularities on the garage surface. Biotic factors affecting coral growth on the garage included fish and invertebrate predation, browsing, and herbivore grazing, competition between adjacent coral colonies, mechanical damage and disease.

The different sides of Cousteau's underwater garage present a set of conditions, almost like a giant "reef ball," that have loose analogues on the reef itself. The geometric symmetry of the domed structure provides gradations of many important biotic ecological factors such as depth, pitch (i.e., slant), orientation, light intensity, current, and turbulence. The dome is raised above the seabed, and water currents are vigorous enough to eliminate significant sedimentation. Biotic factors affecting coral growth on the garage are as follows:

(1) fish and invertebrate predation, browsing, and herbivore grazing
(2) competition between adjacent colonies
(3) competition with, and attack by, sponges
(4) mechanical damage caused by anthropogenic intervention
(5) impact of coral bleaching due to high SSTs
(6) disease

Distinct patterns of coral distribution on the outer surface of the garage were observed. The richest coral growth appeared to be present on side A to B, which faced the prevailing northerly winds and was most exposed to the long-reef current that was usually observed at the garage. This was occasionally one to two knots in strength but normally about half a knot. Its direction was either northerly or southerly, but the garage tended to form an obstruction, and offshoots of the main current swept around the sides, especially on sides B to C and D to A. At times, when strong winds create surface currents, there were small whirlpools at the surface above the garage. Side A to B faced the westerly deep reef face of Shaab Rumi, while side C to D faced the lagoon and shallow reef face (Figure 15.2).

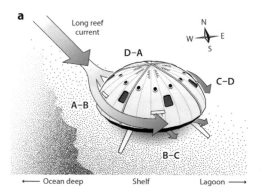 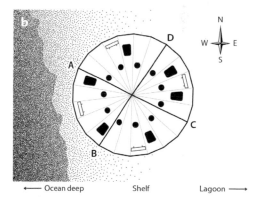

Figure 15.2 Orientation of Cousteau's underwater garage and quadrants studied, relative to the current and deep reef face of Shaab Rumi. Side A to B received most of the incoming current, while side C to D remained relatively protected (as they faced the back reef of Shaab Rumi). Illustrated by Fiona Martin.

© Vine

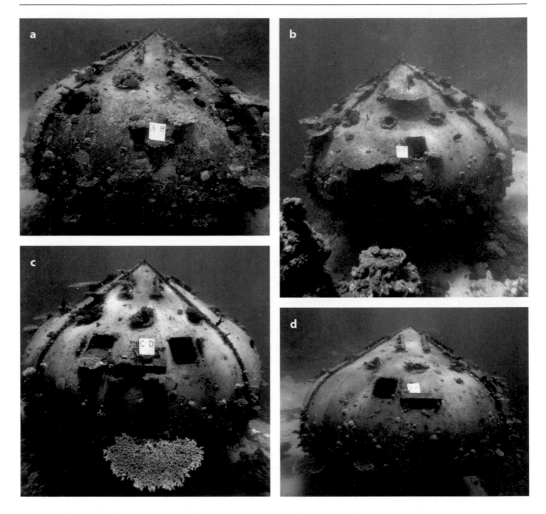

Figure 15.3 (*a*) 1974, side A to B, with coral growth well under way, faces approximately due west; (*b*) 1974, side B to C faces approximately south; (*c*) 1975, side C to D faces approximately east. A table of *Acropora pharaonis* extended 39 cm in length in 10 months. (*d*) 1975, side D to A faces north. Recent studies in Australia have highlighted the importance of table *Acroporas* in reef recovery.

© Vine

The coral species that established themselves on the garage were mainly common species in the region, some with growth rates exceeding most, if not all, previously recorded rates. For example, a fragile table of *A. pharaonis* extended 39 cm in length over 10 months, and a more solid colony of *Porites somaliensis* measured 32 cm across the base, indicating a minimal annual width increment of 2.91 cm.

In a related study, Dr. Juan Carlos Ortiz at AIMS in Australia recently highlighted the influence of *Acropora* tables on the recovery rates of coral reefs after major events. Table corals are incredibly fast growing.

Habitats in exposed reef slopes recover from disturbances at a rate 14 times higher – that's more than two decades faster – when table corals are abundant. Their large, flat plate like shape provides vital protection for large fish in shallow reef areas and serves as a shelter for small fishes, with some species almost entirely dependent on table corals.

Seven genera and 15 species of corals were collected from the garage (Figure 15.5). Considerable care was taken to avoid taking more than one specimen of each species, and several more species may have been present. Some, such as *Tubastraea micranthus*, *Balanophyllia* sp., and *L. corymbosa*, in addition to flourishing soft corals, were present in shaded situations, such as between the double walls of the garage or under the dome.

Figure 15.4 Brain coral. *Platygyra sp.*
© Vine

Plate number	Coral species	Reference number	Relative abundance	Largest colony (April 1975)	Max. annual increment	Max. annual increment (Ting-Ying)
2a	Acanthastrea echinata	12CD	isolated colonies	—	—	1.4 cm
2b	Acropora cytherea	A11	abundant	—	approx. 20 cm	3.2 cm
—	Acropora humilis	AB	quite common	45 × 12 cm	4.1 cm	1.3 cm
2d	Acropora pharaonis	A2 to A7	abundant	153 × 118 cm	approx. 39 cm	—
2e	Cyphastrea micropthalma	3CD	isolated colonies	17 × 14 cm	min. 1.54 cm	0.25 cm
2f	Echinopora lamellosa	4AB	isolated colonies	—	—	2.1 cm
3a	Favia speciosa	1CD, 14CD	abundant	23 × 28 cm	min. 2.54 cm	1.0 cm
3b	Favia pallida	15CD	quite common	—	—	1.0 cm
3c	Goniastrea retiformis	7CD, 10CD	quite common	26 × 6 × 2 cm	min. 2.36 cm	0.55 cm
3d	Leptastrea bottae	6CD	quite common	30 × 32 cm	min. 2.91 cm	0.7 cm
4a	Leptoria phrygia	2CD	isolated colonies	11 × 9 cm	min. 1.00 cm	1.1 cm
4b	Platygyra deadalea	5CD	isolated colonies	27 × 26 cm	min. 2.45 cm	0.5 cm
4c	Platygyra lamellina	13CD	isolated colonies	—	—	0.5 cm
x	Pocillopora favosa	9CD	quite common	27 × 23 cm	min. 2.45 cm	—
4d	Porites somaliensis	16CD	abundant	32 × 30 cm	min. 2.91 cm	—

Key

Isolated colonies   Quite common   Abundant

Figure 15.5 Growth of corals on Cousteau garage.

© Vine

Certain local currents occurred on the garage itself and were significant in their effects on corals. Around the base of the outer wall of the garage, there was usually a weak back-and-forth current created by surges and pressure fluctuations in the domed garage. This effect was more marked at the broken portholes on the upper sections of the garage. Here, a strong oscillating current greatly increased water movement at the edges of the portholes, and the presence of large colonies of *Pocillopora favosa* at these sites, but not elsewhere at that level on the garage, was clearly associated with this factor (Figure 15.6).

Acropora brackets and tables were formed by *Acropora corymbosa* and *A. pharaonis*. Their sizes were measured in April 1975 and again in February 1976 (Figure 15.7). Differences in recorded sizes were attributed to growth, with the proviso that certain corals suffered death, damage, or complete removal during the ten months.

Cousteau reported that an acropore on one of the cables jettisoned from the *MV Rosaldo* had grown to a diameter of almost eight inches (over 20 cm) in four years.

The success or otherwise of corals in this regard influences shifts in dominance between non-reef builders and corals (Figure 15.8). It was apparent on Cousteau's garage, with fresh

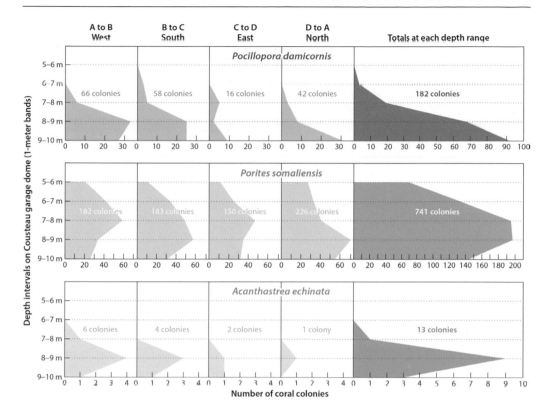

Figure 15.6 The distribution of colonies of *Porites*, *Pocillopora*, and *Acanthastrea* on Cousteau garage.

© Vine

Sector A to B						
Colony code number	Coral species	Length on 15/4/1975	Length on 12/2/1976	Breadth on 15/4/1975	Breadth on 12/2/1976	Length increase	Breadth increase
1	*Acropora cytherea*	22 cm	39 cm	16 cm	28 cm	17 cm	12 cm
2	*Acropora pharaonis*	20 cm	30 cm	30 cm	20 cm	10 cm	6 cm
3	*Acropora pharaonis*	32 cm	dead, missing	30 cm	dead, missing	—	—
4	*Acropora pharaonis*	35 cm	54 cm	25 cm	37 cm	19 cm	12 cm
5	*Acropora pharaonis*	—	—	—	—	—	—
6	*Acropora pharaonis*	21 cm	38 cm	17 cm	30 cm	17 cm	13 cm
7	*Acropora pharaonis*	—	—	—	—	—	—
8	*Acropora pharaonis*	43 cm	58 cm	41 cm	32 cm	15 cm	broken
9	*Acropora pharaonis*	42 cm	45 cm	29 cm	35 cm	3 cm, broken	6 cm, broken
10	*Acropora pharaonis*	60 cm	65 cm	51 cm	53 cm	5 cm	2 cm, broken
11	*Acropora pharaonis*	25 cm	38 cm	21 cm	25 cm	13 cm, dead	4 cm, dead

Figure 15.7 Growth of *Acropora corymbosa* and *A. pharaonis* on Cousteau garage (Continued).

© Vine

12	Acropora pharaonis	25 cm	37 cm	23 cm	—	12 cm	—
13	Acropora pharaonis	89 cm	90 cm	49 cm	70 cm	1 cm	21 cm
14	Acropora pharaonis	46 cm	dead	30 cm	dead	—	—
15	Acropora pharaonis	73 cm	90 cm	40 cm	42 cm	17 cm	2 cm

Sector B to C							
Colony code number	Coral species	Length on 15/4/1975	Length on 12/2/1976	Breadth on 15/4/1975	Breadth on 12/2/1976	Length increase	Breadth increase
20	Acropora pharaonis	29 cm	—	25 cm	—	—	—
21	Acropora pharaonis	—	—	—	—	—	—
22	Acropora pharaonis	20 cm	26 cm, dead	16 cm	25 cm, dead	6 cm, dead	9 cm, dead
23	Acropora pharaonis	53 cm, mostly dead	52 cm	37 cm, mostly dead	40 cm	–1 cm	3 cm
24	Acropora pharaonis	140 cm	150 cm	113 cm	106 cm, broken	10 cm	broken
25	Acropora pharaonis	112 cm	151 cm	59 cm	90 cm	39 cm	31 cm
26	Acropora pharaonis	45 cm	57 cm	43 cm	40 cm, broken	12 cm	broken
27	Acropora pharaonis	23 cm	—	23 cm	—	—	—
28	Acropora pharaonis	56 cm	60cm	44 cm	40 cm, broken	4 cm	–4 cm
29	Acropora pharaonis	29 cm	47 cm	24 cm	30 cm	18 cm	6 cm
30	Acropora humilis	29 cm	36 cm	22 cm	28 cm	7 cm	6 cm

Sector C to D							
Colony code number	Coral species	Length on 15/4/1975	Length on 12/2/1976	Breadth on 15/4/1975	Breadth on 12/2/1976	Length increase	Breadth increase
40	Acropora pharaonis	153 cm	166.5 cm	118 cm	118 cm	13.5 cm	0 cm
41	Acropora pharaonis	—	62 cm	—	71 cm	—	—
42	Acropora pharaonis	—	57 cm	—	29 cm	—	—

Sector D to A							
Colony code number	Coral species	Length on 15/4/1975	Length on 12/2/1976	Breadth on 15/4/1975	Breadth on 12/2/1976	Length increase	Breadth increase
51	Acropora pharaonis	89 cm	111 cm	64 cm	77 cm	22 cm	13 cm
52	Acropora pharaonis	46 cm	50 cm	32 cm	41 cm	4 cm	9 cm
53	Acropora pharaonis	21.5 cm	27 cm	19 cm	22 cm	5.5 cm	3 cm
54	Acropora pharaonis	—	—	—	—	—	—
55	Acropora pharaonis	15.5 cm	—	13 cm	—	—	—
56	Acropora pharaonis	12 cm	dead	11 cm	dead	—	—

Figure 15.7 (Continued)

Figure 15.8 Over the years on Cousteau's garage, biological dominance has shifted from corals to non-reef builders. (*a*) Early photo (1972) showing large *Acropora* tables (© F. Jack Jackson/Alamy). (*b*) Later photo (1990) showing mostly soft corals, sponges, and algae.

larval settlement continuing to occupy vacant surfaces. However, as the most suitable sites became colonised, there were less optimal situations available for the attachment and growth of new coral colonies. Sponges frequently diminish after killing corals, especially during the winter months, allowing algae and invertebrates to settle on the dead skeletons. In some cases, dead colonies broke off and left more space on the surface of the garage for new settlement and growth. As corals were weakened due to various stress factors, they were more susceptible to aggressive competitors such as *Terpios hoshinota*.

Section III

Turf Wars

Unfortunately, a phase shift is taking place towards fleshy macroalgae that can be toxic to coral polyps. The role of seaweeds or macroalgae would be hard to overstate, particularly when considered as habitat-building organisms in biotopes that were once, or still are, dominated by corals. Seaweeds and other macroalgae compete for space on the reef, frequently replacing reef-building corals as the dominant structural reef organisms. They are also a key source of sediments, primary producers, and food for a wide variety of reef organisms (Fong & Paul 2011). Ubiquitous, diverse, and highly adaptable, they are pivotal members of the coral reef ecosystem, frequently influencing outcomes for key faunal groups, from microbial forms to fish and marine mammals (Barott et al. 2012).

Figure III.1 Halimeda algae covering a bleached coral reef. *Halimeda* is typically a recoloniser on dead coral.

DOI: 10.1201/9781003335795-18

Fleshy macroalgae are among the first to take possession of newly exposed surfaces created by physical disturbances such as storms, dredging, CoTS, or coral bleaching, with the algae generally "stealing a march" on reef-building corals, their competitors, for space on the reef. The result can be a reduction in growth rates of both corals and macroalgae, reducing fecundity or even triggering mortality in affected coral colonies. Nicola Foster (2005) refers to the ebb-and-flow pattern of growth and decline on reefs as "patch dynamics," which she defines as the net sum of colonisation rate, growth rate, and extinction rate on coral reefs.

Today's coral reef/algae communities tend to be dominated by encrusting algae, CCA, and turf-forming filamentous algae that are more compatible with healthy reefs than the brown, fleshy macroalgae that have taken over on compromised reefs in recent years. Numerous biologists have recently drawn attention to this phase shift that is displacing corals. The process frequently follows the loss of herbivores due to overfishing (Ceccarelli et al. 2006), pollution, sedimentation, disease (in corals and CCA) (Quéré & Nugues 2015), invasive species, and climate change (Florian et al. 2021). It is happening on most coral reefs and is difficult to prevent or reverse. What were until recently diverse coral reefs, supporting rich communities of fish, invertebrates, sea turtles, and marine mammals, have turned, in parts of the Caribbean, Pacific, Indian Ocean, and Red Sea, into resilient algal reefs, replacing their coral-dominated predecessors.

Chapter 16

Ecological Impact of CCA, Filamentous Algae, and Grazing by Fish

The role of crustose coralline algae (CCA) in reef development is generally positive (Chisholm 2003), but not always so. Aeby et al. (2017) studied diseases affecting CCA and their significance for coral larvae. They documented three CCA diseases – coralline lethal orange disease, coralline white syndrome (CWS), and coralline white band disease – among which CWS (white patch disease) was the most common in the Red Sea and could be accompanied by important downstream effects such as reduced survivorship of coral larvae (Quéré & Nugues 2015).

The author's interest in CCA's role in reef development was triggered by settlement studies in the early 1970s (Vine 1974), when it became apparent that CCA cemented and bound loose coral rubble in areas of reef where herbivores were not excluded by aggressive damselfish (Figure 16.1). The farmer damselfish, *Stegastes lividus*, actively defends inhabited patches of algal turf on shallow reefs in the Red Sea. In doing so, it excludes herbivorous fish, which normally crop filamentous red algae (e.g., *Polysiphonia*). The aggressive pomacentrids browse on the upright axes of selected filamentous rhodophytes that grow within their territories, and they actively weed out indigestible species. The latter tend to comprise species that the damselfish are unable to digest due to a lack of suitable enzymes (carbohydrase) or masticatory organs. The result is a well-maintained monoculture that would otherwise be susceptible to grazing by herbivores, as well as competition from other algae, rendering fish and turf algae mutually dependent on each other (Hata et al. 2012).

CCA can suppress green algae, having potentially dramatic impacts on reef recovery. In addition to its role in cementing the reef into a solid structure, it also helps limit the excessive growth of algae such as *Ulva*. It has been shown that the absence of CCA leads to an overgrowth of such algae and the likely death of smothered corals. CCA limits the local abundance of already established macroalgae by reducing both their growth rate and their recruitment success. Conversely, crustose coralline species induce settlement and metamorphosis in a large number of scleractinian corals.

As mentioned earlier, the main stimulus for competition is the need to find space to settle and grow. The impact of turf algae on settlement by corals and other sessile organisms is described by Hata and colleagues (e.g., Hata 2002; Hata & Kato 2004). In general, as we have seen, turf algae grow so quickly and at such a high density that they take over any space that is not regularly grazed or scraped by a variety of herbivores, such as fish, sea urchins, starfish, molluscs, and crustaceans.

While filamentous algae that is ubiquitous on shallow reefs tends to exclude coral larvae from settling and occupying exposed surfaces, its real impact may be more nuanced. Speare et al. (2019) showed how accumulations of sediment among the algal fronds can strike a fatal blow to coral planula larvae. While these can cope with just the coating of turf algae, they are less

DOI: 10.1201/9781003335795-19

EFFECTS OF HERBIVOROUS FISH ON ALGAE, CCA, AND CORAL GROWTH

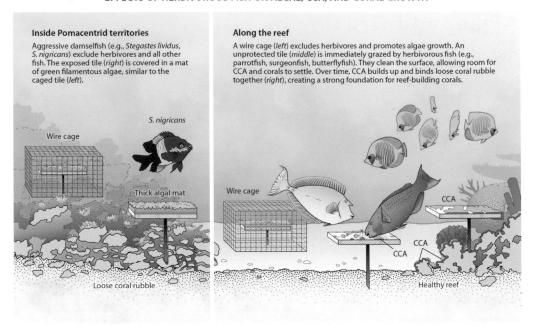

Inside Pomacentrid territories

Aggressive damselfish (e.g., *Stegastes lividus*, *S. nigricans*) exclude herbivores and all other fish. The exposed tile (*right*) is covered in a mat of green filamentous algae, similar to the caged tile (*left*).

Along the reef

A wire cage (*left*) excludes herbivores and promotes algae growth. An unprotected tile (*middle*) is immediately grazed by herbivorous fish (e.g., parrotfish, surgeonfish, butterflyfish). They clean the surface, allowing room for CCA and corals to settle. Over time, CCA builds up and binds loose coral rubble together (*right*), creating a strong foundation for reef-building corals.

Figure 16.1 An experiment to investigate the role of CCA, herbivory, fish grazing, filamentous algae, and larval settlement on coral reefs. Pomacentrids act as "farmers," selectively removing indigestible algae and protecting the growth of fine, filamentous algae. Thick algal mats prevent the settlement of binding CCA and corals, eventually leading to mounds of loose coral rubble. Removing herbivores by other means – such as overfishing or pollution – may also leave algal growth unchecked, reducing space for CCA and corals to settle. The survival of reefs in the future may depend on the presence of a healthy population of fish. Illustration by Fiona Martin.

© Vine

resilient in the face of sediments that clog their fine tentacles: "sediment significantly impedes coral settlement, reducing settlement 10- and 13-fold for *A. palmata* and *O. faveolata*, respectively, compared to turf algae alone" (Speare et al. 2019).

Bathroom tiles were attached to Harvey Reef. After 2 and 4 weeks, the tiles were removed from the reef, and any settled algae were scraped off, dried, and weighed (Figure 16.2). There were four categories of tiles:

- Tiles exposed on the reef, that is, control samples
- Tiles protected from reef grazers by wire cages
- Tiles protected from other reef grazers by *S. lividus* damselfish
- Tiles protected from reef grazers by *A. sohal* surgeonfish

The results of our scraping and weighing of filamentous algae were both affirmatory and unexpected. They demonstrate some basic tenets of coral reef ecology (Figure 16.1).

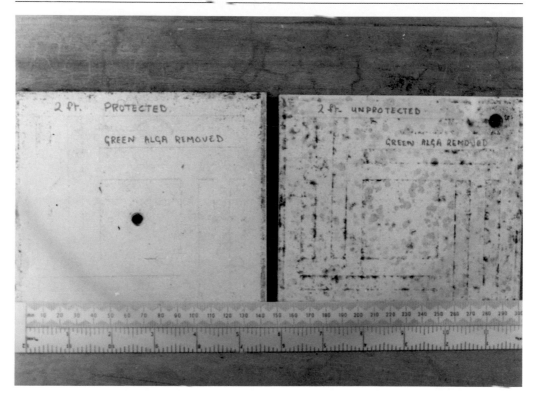

Figure 16.2 Two experimental tiles exposed to different conditions on the shallow reef in Sudan show the important role of "farmer fish" cropping filamentous algae and making space for the settlement of CCA.

© Vine

1   Tiles that were protected by cages had noticeably more algae than did unprotected tiles in the same location, providing that aggressive fish were not excluding algal browsers. This result was expected and due to the absence of normal grazing on the protected tiles. The situation was analogous to what would happen if all the fish were removed, leaving nothing to graze down the filamentous alga. Reduction of reef herbivores due to overfishing recreates these conditions, resulting in the proliferation of filamentous algae and the knock-on consequences that ensue.

2   A comparison between protected and unprotected tiles placed within a farming fish territory, such as those created by *S. lividus* or *A. sohal*, shows little difference between the protected and unprotected tiles. This is because the aggressive fish are serving the same purpose as the wire netting cages: excluding herbivorous fish. Consequently, the dry weights of algae on the two plates were almost identical.

3   What we saw after scraping algae off the plates proved to be a surprise. While the protected tiles had the most growth of filamentous algae, they were otherwise bare of any other settled organisms. This is clear from the photograph of two tiles that was taken at the time (Figure 16.2).

4   In sharp contrast, the tiles that had been left unprotected experienced vigorous, prolific settlement and growth of pink crustose coralline algae (CCA) that were prominent on the tiles after removal of the filamentous algae. The experiment thus demonstrated that a healthy population of herbivores and browsers, scraping at the reef surface, created space for CCA to settle and grow.

We were impressed by the far-reaching implications of these findings for the region's coral reefs, as well as those around the world. Other researchers were beginning to reach similar conclusions, and the importance of coral/algae/fish dynamics, resulting in phase shifts from corals to macroalgae, has since become a pivotal element of reef ecology. These findings were originally reported in the journal *Marine Biology* (Vine 1974), and subsequent studies are discussed in Vine (2019).

Implications of the algal-coral phase shift include, but are not limited to, the following features:

1   Loss of herbivores (fish, echinoids, and other invertebrates) results in overgrowth of algae (both filamentous and macroalgae), killing corals.
2   Overfishing tips the balance in favour of a takeover by algae on coral reefs.
3   Farming species, such as *S. lividus and A. sohal*, can have a deleterious impact on corals, promoting algae over corals.

Figure 16.3 Acanthurus sohal defends a crop of filamentous algae that has taken over dead coral, excluding other herbivores.

© Vine

4   Damage to corals caused by violent physical disturbances requires protection of fish species to promote regrowth and reef recovery.

5   Crustose coralline algae (CCA) helps to provide rigidity to coral reefs, cementing loose fragments together.

6   CCA is a vital element of healthy coral reefs and competes successfully for space on the reef, establishing itself on surfaces grazed by herbivores.

7   CCA is a favoured settlement surface for reef corals (Tebben et al. 2015).

8   CCA is susceptible to diseases in warm, shallow waters and remains a key species in reef development and recovery.

9   CCA provides some essential ingredients for healthy coral reefs, including a number of microorganisms that help create suitable conditions for coral larvae to settle. These may be rendered ineffective by ocean acidification (Webster et al. 2013).

10   CCA may be attacked by physical factors, diseases, and a number of antagonists. Loss of healthy CCA is likely to be accompanied by an upsurge in fleshy macroalgae, which may themselves exude compounds that are toxic towards corals – thus establishing a cascading effect and rapid deterioration of the reef environment.

Chapter 17

Sponging on Coral

Reversing the downward trends on coral reefs may not even be possible, but if there is ever going to be progress in this field, it is necessary to recognise the scientific barriers that must be crossed (González-Rivero et al. 2011).

While diving in the central Red Sea in 1972, I became fascinated, not so much by its profusion of life as by its alarming rate of death. Initially distracted by the vast array of colourful fish and corals that adorned the reef, I had become aware of an insidious, paper-thin tissue that was casting a deathly pall over all that came within its reach. On closer study, I concluded that this amorphous organism was wreaking havoc over vast areas of coral reefs, including some of the Red Sea's reefs.

My best guess was that this species was a previously undescribed, encrusting sponge that had remarkable powers to poison and kill reef-building corals. It was happening so quickly that one could almost watch it grow day by day. Somewhat inconspicuous by virtue of its dull colour and reef-hugging profile (Elliott 2016), there was a recognisable barrier at the undulating edge of the encrusting sponge, where it came in touch with the threatened live corals. Was this a new phenomenon or one that was just gaining notoriety as more and more divers explored the reefs? It was an unsolved mystery, with the only previous record that I could track down being an account of a dive in the Indian Ocean by Jacques Cousteau.

I was living in close proximity to the reefs and was able to return, time after time, to the same coral colonies and record their decay. The process was repeated with concerning predictability. First, the toxic blanket established a "bridge head" over a small section of live coral, and then it extended itself out over the undulating surfaces of the coral polyps, often covering more than 10 square metres in a month or two, transforming the vigorous ecosystem of the reef to a monotonous grey-black hue, and killing any coral that it covered. Wherever the sponge occurred, the visible effect could be quite dramatic, reminiscent of the damage caused by CoTS. The colour, fish, and fascinating invertebrates of the reef were gone, bringing a growing sense of despair. A short description was published in 1977, in the newly established *Journal of the Saudi Arabian Natural History Society* (Vine & Head 1977).

My search for information on the killer sponge was particularly focused on desk and field-work in 1975 and 1976. I corresponded with a number of invertebrate museum curators and experts on coral reefs, including F. R. Fosberg (January 1976), Klaus Reutzler at the Smithsonian Institution (21 October, 1975), Henry Reiswig at Redpath Museum (1975), and others. The general consensus was that up until 1975, few had heard of the recently recognised threat to coral reefs from encrusting sponges.

It is fascinating now to look back at the slow pace of communications at that time, compared with today's split-second Internet and its instant access to knowledge. However, the wheels of

DOI: 10.1201/9781003335795-20

academia still turned, albeit at a slower pace. A letter dated 21 January 1976, from the Professor of Biology at Yale University, Professor Willard Hartman finally brought the answer that I had been seeking. He wrote:

> Your letter to Bill Newman about the coral-killing sponge has just reached me by way of Henry Reiswig. I suspect that he has written to you already about the blackish encrusting sponge that we both observed in Guam overgrowing living corals. A graduate student at the University of Guam has since published a note on the sponge which Dr. Berquist and I had independently identified as a new species of *Terpios*.

Dr. Reiswig had indeed written to me in response to my observations on the enigmatic sponge, mentioning its encrusting form and its devastating impact. He commented as follows:

> In August 1971, Dr Willard Hartman (Yale U) and I made a field trip to Guam/Saipan to inspect the sponge fauna of submarine caves. Off a small islet on the east coast of Guam, Anae Is., I noted extensive dead coral areas and a thin encrusting sponge which was responsible for at least part or all of the damage. His field notes from Anae Is. dated, 9 August 1971, read:

>> Purple-grey killer sponge all over in patches up to 10 meters across; is killing and covering living coral at the front of direct contact; without a clear dead area between sponge and coral.

Dismissing the possible cause of dead coral being connected to CoTS, he added:

> I could not attribute the local devastation to *Acanthaster* in view of the obvious interactions seen to be taking place and the extent of cover attained by this sponge.

He added a more general comment on the geographic spread of *Terpios* as he had seen it.

> It is noteworthy that the alleged sponge kill at Anae Is. was a local phenomenon. Diving a mile or two north and south of that location turned up no other instances of this interaction.

While the earliest paper referring to this sponge was that of PG Bryan in 1973 (in the journal *Micronesica*) it was not until 1993, 20 years after my observations in Sudan and coinciding with the published account of the spread of a new *Terpios* species in Guam, that a full account was published (Rützler et al 1993).

My own comments on how the killer sponge was affecting corals that were growing on Cousteau's underwater garage were as follows:

> Between visits [to Cousteau's garage] made in April 1975 and February 1976 a significant number of corals died or suffered damage. Of the bracket and table forms of *Acropora*, 44 percent suffered some form of damage and the death toll was 25 percent of the total. All the colonies that were noted as being in poor condition or partially dead during April 1975 had completely died and in some cases broken off from the garage by February 1976.

The most significant cause of mortality was penetration or surface attack by sponges. There were at least four species actively involved in this process: an unidentified green species (possibly

Dysidea sp.), a grey encrusting species (*Terpios* sp.), an orange-coloured encrusting species (similar to *Spirastrella* sp.) and a red boring sponge. On side C to D of the garage, above and around the large table of *Acropora pharaonis*, it was possible to see the lateral extension by *Terpios* during the ten-month period between visits. This encompassed a number of coral colonies that had previously been living (in late 1975) but were dead by February 1976. Certain colonies, such as *Favia speciosa*, demonstrated notable resistance to attack and remained as isolated living coral colonies surrounded by encrusting grey sponge.

The phenomenon of corals being killed by encrusting sponges was becoming more and more noticeable (Pang 1973) and it was apparent that, following active periods of sponge extension and killing corals, particularly during summer months in the Red Sea, there may be regression and even complete disappearance of the sponge. In some cases, it was shown that corals that were partially killed by such sponges (*Terpios hoshinota*) may recover following the retreat of the sponge. It has been reported, however, that coral planulae do not select as settlement substrata previous *Terpios hoshinota* killed corals.

Rützler and Muzic's original description matched my own notes on the killer sponge as it appeared in the Red Sea, with the notable difference of having a greater depth range: "It is recognised by its extensive greyish to blackish encrustations on coral, distinctive lobed tylostyle spicules, and an association with abundant, large, unicellular cyanobacteria of the *Aphanocaps raspaigella* type." They described how the sponge aggressively competes for space by killing and overgrowing live coral and is responsible for the demise of large reef areas, particularly in pollution-stressed zones near shore. Given that its symbiotic zoocyanellae make up half or more of the sponge tissue, it seems likely that the physiological behaviour of the sponge is closely aligned with that of the zoocyanellae.

A typical encounter with "black disease" as these infestations by *Terpios hoshinota* were now called, was that described at Yongxing Island, the largest reef island of the Xisha Islands in the South China Sea. Prior to 2002, coral reefs at its north end were flourishing, but a sharp decline took place in the following few years, so that surveys in 2008 and 2010 revealed coral mortality on a large scale, accompanied by the encrusting blackish sponge mat (identified as *Terpios hoshinota*). As sea temperatures fell, coral growth became more vigorous, and "black disease" receded. Later, as spring and summer reheated the ocean, corals once again succumbed to the sponge's toxic presence, dying as they were overgrown (Elliott 2016).

It became clear that this is an aggressive species that kills corals with apparent impunity. It grows much faster in linear dimensions than corals – twice as fast in fact – and hence has the propensity to overgrow corals, shading them from sunlight (which they depend on for photosynthesis) and ultimately causing their death, often on a scale of tens of square metres.

Given the sponge's indirect dependence on photosynthesis (by cyanobacteria), it is not surprising that the seemingly inexorable advance of the sponge's extension over live stony corals can be slowed down by reducing light intensity with artificial shading. During the cooler months of January to April in the central Red Sea, I noticed how the sponge slowed down its growth as the seawater temperature fell, coinciding with a general boost to corals, which in winter were capable of covering over the sponge and killing it. Unfortunately, as the sea temperature rose back towards summer levels, the sponge responded with a re-possession of territory, once more smothering and killing the coral. I observed this back-and-forth battle between *Terpios* and corals for several cycles.

Following its observed presence in Guam (Bryan 1973) and the Sudanese Red Sea in 1975 (Vine P, pers.obs.), it was reported in tropical and subtropical seas where corals normally thrive including the Indo-Pacific, Red Sea, Seychelles, American Samoa, Polynesia, Micronesia,

Melanesia, the Philippines, Japan, Taiwan, Australia's Great Barrier Reef, Indonesia, Maldives, India, Mauritius, and the Atlantic.

But there is a larger aspect to the destruction caused by *Terpios hoshinota*. The cyanobacteria that make up much of their cytoplasm are also valuable sources of bioactive substances that are the basis of new pharmaceuticals with antiviral, antibacterial, antifungal, and anti-cancer activities. They are also a key element of new hydrogen power, considered a very promising source of alternative energy. Cyanobacteria are also used in aquaculture, wastewater treatment, food, fertilisers, and production of secondary metabolites including vitamins and enzymes. Furthermore, oil-polluted sites, such as those generated by wars or acts of terrorism, generally develop rich populations of cyanobacteria that degrade, and thus help to clean up, the oil.

Coral reef survival is influenced by zoogeographical conditions in terms of tolerances to temperatures, salinity, acidity, and other anthropological influences. A key issue will be increasing acidity, leading to dissolution of calcium carbonate – a fundamental structural building block of life in the oceans.

In the case of corals versus algae, there is an extended conflict influenced by fluctuating conditions, with positive or negative consequences for the combatants, such as the disappearance of herbivorous fish caused by overfishing. Loh et al. (2015) demonstrated the indirect effects of overfishing on Caribbean reefs in the form of sponges overgrowing reef-building corals. Other triggers for boosting algae may involve sedimentary deposits (Speare et al. 2019), enhanced nutrients, storm damage, or coral bleaching, all of which may be caused by climate change.

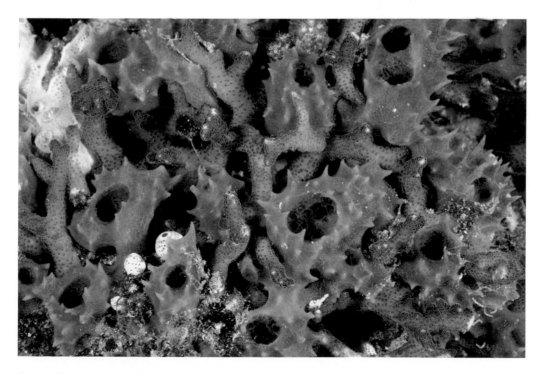

Figure 17.1 Sponges (*Callispongis* sp.) encrusting hard coral.
© Images & Stories/Alamy

Such events can be compared with the on-land equivalent of removing herbivorous farm animals (e.g., grazing cows, goats, horses, or donkeys) from a meadow full of wildflowers. In the absence of grazers and browsers, the most aggressive grasses will soon dominate the sward, transforming diverse, rich natural habitats into monospecific grassland, which supports greatly reduced numbers of insects, birds, and other faunal groups.

Corals, corallimorpharians, algae, and sponges all incorporate photosynthesis into their metabolic regimes, and light is a significant factor, along with the deployment of toxins, in their growth dynamics. Any organism capable of blocking the light from reaching reef-building corals has the potential to kill the corals, making space for a takeover by it or another species. The complex interrelations that characterise the coral–algae–sponge interfaces have been the subject of close study since the early 1970s, when such interfaces were first noticed by field biologists able, thanks to SCUBA, to monitor the growth and decay of these three groups over prolonged periods.

Terpios was originally described by Keller (1889) from a specimen, collected near Suakin, under the name *T. viridis*. It was found in the *Stylophora* zone and studied as part of Vine's survey on Sudanese reefs (Vine and Vine 1980). A series of observations clearly demonstrated how rapidly this sponge can spread and kill living corals. One extensive matting of the sponge extended its coverage to an area of 400 sq m during the summer months in the depth range of 15–20 m along a 40-m length of reef on which all smothered corals were killed. Antonius (1988a, 1988b) summarised the spread of *T. hoshinota* in the eastern Red Sea. Meanwhile,

Figure 17.2 Live coral being smothered by the coral-killing sponge *Terpios* cf. *hoshinota*.
© Vine

there were reports from Guam that *Terpios hoshinota* was killing large sections of *Porites*. The coral *Galaxea* was most likely to "fight back" during the cooler winter months. The impact of shifts from coral-dominated to algal-dominated habitats varies in their significance, but there is generally a strongly negative effect if the macroalgae are left unchecked (see Wild et al. 2014). However, not all macroalgae are deleterious to reef growth, and CCA may enhance coral settlement and early survival.

The impact of macroalgae on coral reefs can be devastating, including "reduced fecundity and larval survival, pre-emption of space for settlement, abrasion or overgrowth of recruits, sloughing or dislodgement of recruits settled on crustose algae, and changes to habitat conditions." For a discussion of space on coral reefs, see Benayahu and Loya (1981). Barott et al. (2012) recognised three types of interaction between corals and algae: coral damaging algae, algae damaging coral, and "stalemate" in which there is no clear winner. The outcomes of such competition can be affected by a variety of natural and anthropogenic factors, often influenced by seasonal, diurnal or global warming associated with sea temperature rise. These environmental triggers may be critical for development and survival of coral reefs. They occur in a variety of ways, often involving shifts in the associated bacterial communities, including changes to the interactive zones where pathogens and virulence genes come into play.

An obvious question with regard to the impact of fleshy macroalgae on live corals is which comes first: a breakdown in reef corals, allowing algae to settle and colonise available spaces, or an attack on corals by fleshy macroalgae, killing the corals and creating surfaces for algal growth and a shift from coral-dominated to algae-dominated reefs? The answer was eloquently articulated by Smith et al. (2006) in a paper entitled "Indirect Effects of Algae on Coral: Algae-Mediated, Microbe-Induced Coral Mortality," in which they describe an experiment whereby they placed algae and coral in chambers, separated by an ultra-fine filter. All the corals died. However, when the water was treated with a broad-spectrum antibiotic, mortality was entirely prevented. The authors suggest that a "loop effect" is likely to be created whereby "human impacts increase and algae become more abundant leading to release of compounds that enhance microbial activity on live coral surfaces, causing mortality of corals and further algal growth."

As the years pass and coral reefs in the central and northern Red Sea and Gulf of Aqaba show steady signs of degradation, it is increasingly important to establish benchmarks against which this deterioration can be measured. Rinkevich (2005) puts it very clearly: "What Do We Know about Eilat (Red Sea) Reef Degradation? A Critical Examination of Published Literature." The same could be asked of other key areas of the Red Sea. Riegl et al. (2012, 2013) were no less emphatic when describing coral population trajectories over two decades in the Red Sea. One-third of coral reefs distant from human population were sampled in the Red Sea and "showed degradation by predator outbreaks (crown-of-thorns-starfish= CoTS) observed in all regions in all years) or bleaching."

Aswani and colleagues (2015) list the most important steps that need to be undertaken in order to give corals a fighting chance as: (1) enhancing the case for reef conservation and management, (2) dealing with local stressors on reefs, (3) addressing global climate change impacts, and (4) reviewing various approaches to the governance of coral reefs (Fidelman et al. 2019).

Section IV

Underwater Recording

In almost all cases, observations were made while SCUBA diving. Waterproof note paper (Ascot, NCR Appleton Papers Division) was used for sketching the reef face, together with a diving depth gauge for recording depths. Descriptions were thus made while underwater at the location being studied. Rotenone and quinaldine were used for the collection of fishes in Suakin Harbour and on the Port Sudan fringing reef. Pronoxfish was used at several sites for the collection of invertebrates.

The Shaab Rumi underwater garage (located north of Sanganeb) was visited on three occasions, and more than 30 hours were spent measuring and photographing corals growing on this iconic structure. A Nikonos camera with a close-up lens or a Rollei SLR66 camera with

Figure IV.1 The picture shows an aerial view of the Cambridge Coral Starfish Research Group platform at Harvey Reef. It was taken by remote-controlled fixed-wing aircraft.

© Vine

DOI: 10.1201/9781003335795-21

an underwater housing (manufactured by Wolf Koehler) was employed. Underwater measurements to the nearest centimetre were made with a tape measure and ruler. British Admiralty Charts were used for locating positions.

Survey sites were selected with a view to demonstrating the variability of reefs in the region. Thus, locations were chosen from inside sheltered harbours, on more exposed fringing, patch, and barrier reefs, and at the offshore lighthouse reef at Sanganeb atoll, together with the Conshelf

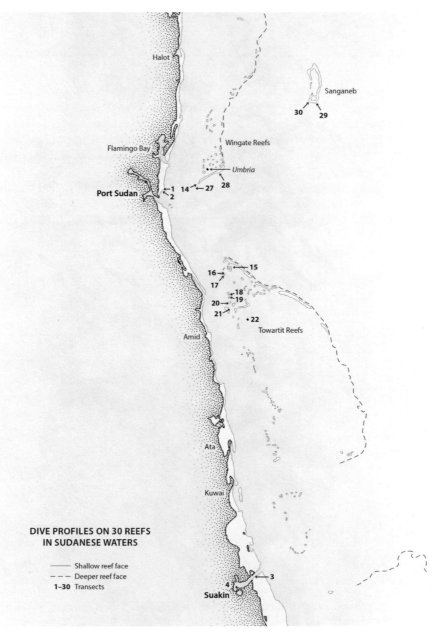

Figure IV.2 Map of Port Sudan and locations of 30 dive transects nearby. Bold numbers mark the locations of transects covered in the following chapters.

© Vine

2 underwater garage. The selective and limited extent of site surveys imposed an acknowledged bias on the data collected. The intention was to present a preliminary description of the broad range of reef forms (Figure IV.3), rather than a classification of reefs, and to thus suggest areas where more intensive research might be concentrated.

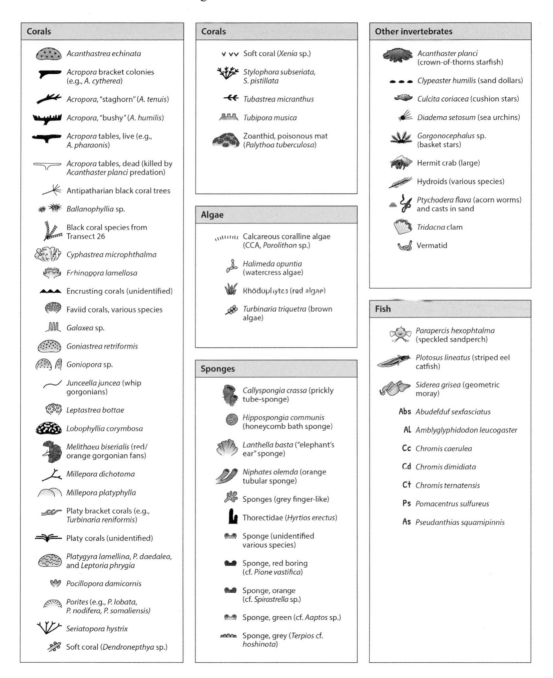

Figure IV.3 Key to biological symbols used in profile drawings of dive transects. Using simple symbols allowed for efficient recording of complex ecosystems.

© Vine

Chapter 18

Port Sudan

Transect 1

Location: Approximately 1 km north of the northern point of the Port Sudan Harbour entrance. The reef changes direction at this promontory so that the reef is more exposed to prevailing winds to the north and less exposed between it and the harbour entrance.

Morphology: The reef crest is deeply indented, and the shallow reef edge forms a spur-and-groove system. Sneh and Friedman (1980) discuss this pattern of growth. The shallow reef face is steep and sometimes undercut, especially under *Porites* colonies, which form the crest. The terrace at 8–9 m is rubble-strewn. A small pinnacle rises to 6.5 m from 10 m. Outside this pinnacle is the deeper reef face, extending to 26 m. There is a narrow shelf at 26–27 m before a gradually steepening reef face that extends to 46 m. The terrace below is composed of fine white sand with isolated rocky outcrops.

General ecological notes: There were marked faunal differences between the two reef sections north and south of the promontory (compare this with Transect 2 below). The more exposed northern section had more prolific growth of hermatypic corals and a less acute reef face. To the south of the promontory (e.g., Transect 2), the reef face was almost vertical or overhanging. Scleractinian corals were replaced by a rich growth of orange gorgonian sea fans (*Melithaea biserialis*), pink fleshy soft corals (*Dendronephthya*), and, at greater depths, black corals (antipatharians). The transect was at the promontory, where an abundant food supply nourished coral growth. The reef lies in line with the long-shore current that delivers plankton to reef inhabitants. On several SCUBA dives at this location, we collected corals down to 80 m. A notable feature of this location was the apparent scarcity of marine life, including fish life below 30 m – a marked contrast with the great abundance and diversity of life in the upper 10–15 m.

Corals: The reef crest was dominated by *Pocillopora* and *Porites*. There was a zone of *Galaxea* at around 2 m. Large but dead tables of *Acropora* occurred on a rubble-strewn terrace at 10 m. The shallow reef face was in a spur-and-groove system, with numerous corals on spurs but few in sheltered grooves where an encrusting pink CCA (cf. *Porolithon*) is frequently observed. Fans of *Millepora dichotoma* at 2–10 m depth confronted a long-reef current. *Balanophyllia* was common in the crevices of the pinnacle, which rose to 6.5 m from 10 m. *Pocillopora* was also abundant on the pinnacle. The deep reef face, extending to 26 m, had several large *Acropora* tables and other corals. These were sparsely distributed, and *Xenia* predominated with various sponges, some of which, including corals smothered by *Terpios hoshinota*.

Fish: Fifty-two species were recorded at this transect, only eight of which were present below 25 m.

DOI: 10.1201/9781003335795-22

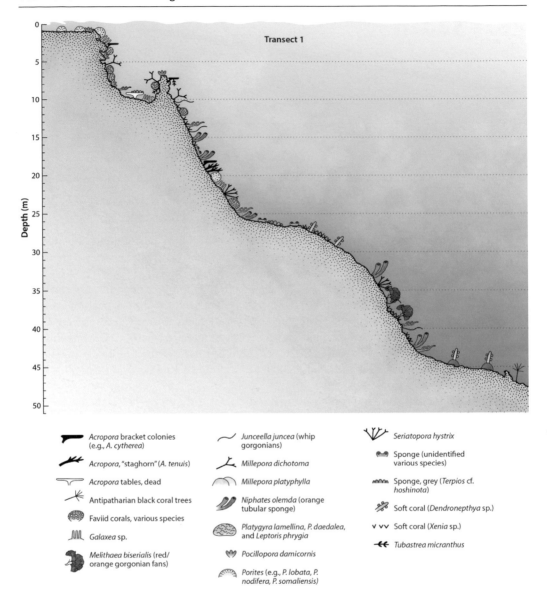

Depth (m)

Transect 1

Acropora bracket colonies (e.g., *A. cytherea*)

Acropora, "staghorn" (*A. tenuis*)

Acropora tables, dead

Antipatharian black coral trees

Faviid corals, various species

Galaxea sp.

Melithaea biserialis (red/orange gorgonian fans)

Junceella juncea (whip gorgonians)

Millepora dichotoma

Millepora platyphylla

Niphates olemda (orange tubular sponge)

Platygyra lamellina, P. daedalea, and *Leptoris phrygia*

Pocillopora damicornis

Porites (e.g., *P. lobata, P. nodifera, P. somaliensis*)

Seriatopora hystrix

Sponge (unidentified various species)

Sponge, grey (*Terpios* cf. *hoshinota*)

Soft coral (*Dendronepthya* sp.)

Soft coral (*Xenia* sp.)

Tubastrea micranthus

Figure 18.1 Transect 1.
© Vine

Transect 2

Location: Approximately 100 m south of Transect 1, on the Port Sudan fringing reef.

Morphology: There is an almost vertical cliff face from the reef crest to about 35 m. The reef top comprised consolidated coral rock, while the reef crest had numerous deep crevices. The reef face was undercut at 3 m, and the relatively sheltered aspect of the reef was reflected by an accumulation of sediment among the crevices. At 10 m, the reef face was steep and undercut in places, with numerous gorgonians and *Dendronephthya* suspended from overhangs.

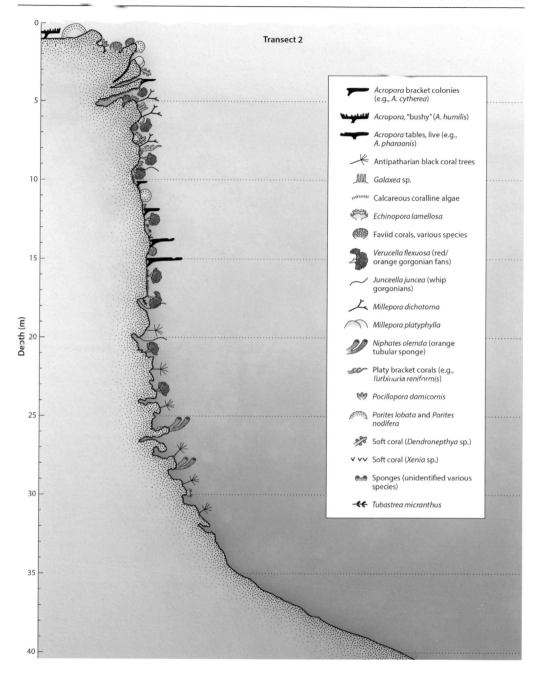

Transect 2

	Legend
	Acropora bracket colonies (e.g., *A. cytherea*)
	Acropora, "bushy" (*A. humilis*)
	Acropora tables, live (e.g., *A. pharaonis*)
	Antipatharian black coral trees
	Galaxea sp.
	Calcareous coralline algae
	Echinopora lamellosa
	Faviid corals, various species
	Verucella flexuosa (red/orange gorgonian fans)
	Junceella juncea (whip gorgonians)
	Millepora dichotoma
	Millepora platyphylla
	Niphates olemda (orange tubular sponge)
	Platy bracket corals (e.g., *Turbinaria reniformis*)
	Pocillopora damicornis
	Porites lobata and *Porites nodifera*
	Soft coral (*Dendronepthya* sp.)
	Soft coral (*Xenia* sp.)
	Sponges (unidentified various species)
	Tubastrea micranthus

Figure 18.2 Transect 2.

© Vine

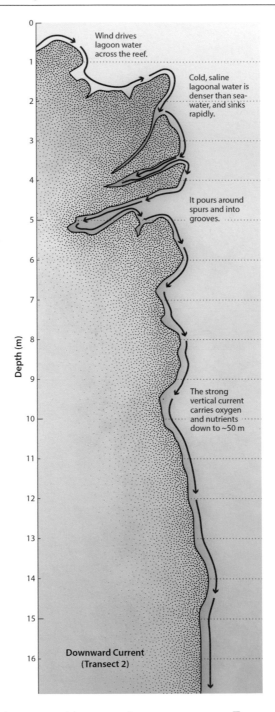

Figure 18.3 Downward current with spur-and-groove system at Transect 2.
© Vine

The spur-and-groove system (see Sneh & Friedman 1980; da Silva et al. 2020) extended vertically from the crest to the base of the reef at 35 m. The series of spurs and grooves influenced currents flowing close to the reef face. During strong north and northwest winds, the reef edge at this point is in line with wind drift from the lagoon. Lagoonal water becomes colder (with latent heat loss due to evaporation in shallows) and more saline and is wind-driven across the reef. On winter days with a strong northerly wind, there was notable mixing of water at the reef crest. Since colder, saline lagoonal water is denser than open seawater, it sinks rapidly upon meeting seawater at the reef edge. The first place where this mixing takes place is above the vertically indented grooves. At these points, heavy lagoonal water poured into the grooves and rapidly sank.

This downward current, extending along a substantial section of reef face, is strongest in the grooves and least powerful about a metre out from the spurs. Observations based on the angle of downward deflection of a (floating) pencil on a string showed that the current is strongest closest to the reef face and diminishes to an undetectable movement at one or two metres away from the steeply sloping reef face. This strong downward movement of water was observed down to 50 m, where its velocity was undiminished. Such movement appeared to extend considerably beyond that depth and is significant in carrying shallow oxygen-saturated water, rich in plankton and burdened with sediment and debris, towards the mesophotic level. Since it is, however, an intermittent phenomenon based on the variable strengths and directions of winds, its influence

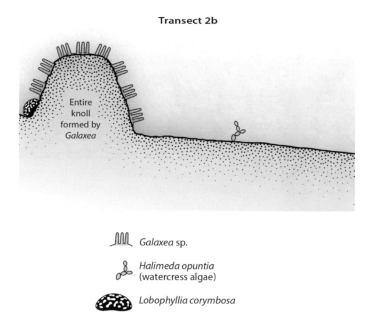

Transect 2b

Entire knoll formed by *Galaxea*

Galaxea sp.

Halimeda opuntia (watercress algae)

Lobophyllia corymbosa

Figure 18.4 Transect 2b.
© Vine

on fauna is difficult to quantify. However, there is usually a scarcity of fish below about 20 m, whereas, during these down-current periods, a large number of typically shallow-water fish occur as deep as 35 m.

Corals: *Pocillopora* sp., bushy acropores, and *Platygyra* occupy the reef top. At 5 m, a large colony of *Galaxea* (see Transect 2b) provided attachment surfaces for several large fans of *Millepora dichotoma* that were facing long-reef currents. In addition, there were brackets of *Acropora cytherea*, several faviidae, and the soft coral *Dendronephthya savignyi* hanging from overhangs, together with small gorgonian fans.

At 8 m, the steep reef face had *Millepora dichotoma*, *Echinopora*, *Pocillopora*, *Porites*, *Fungia*, and various faviidae. At 10 m, the face is very steep and undercut in places, favouring *Verucella flexuosa* and *Dendronephthya*. Below 10 m, the coral growth is noticeably reduced, although there are occasional outcrops of *Porites* and brackets of *A. cytherea*. At 15 m, there were several larger tables of *A. pharaonis*.

Fish: Sixty species were recorded at the transect, of which 21 were present below 25 m. At the time of our observations (January 1976), there was a high zooplankton density, particularly two species of small swimming pteropods. These were providing a rich food supply for many fish, including a large number of species that are not, under normal circumstances, plankton feeders. It was remarkable to watch species normally living closely associated with the reef, such as parrotfish, wrasse, surgeonfish, and triggerfish feeding in mid-water on these pteropods. The following is a list of species that, under these intermediate conditions, were observed to feed on pteropod plankton: *Sardinella*, *P. squamipinnis*, *Pomacanthus asfur*, *Pygoplites diacanthus*, *Chaetodon auriga*, *Chaetodon austriacus*, *Chaetodon fasciatus*, *Heniochus intermedius*, *Acanthurus gahhm*, *Acanthurus xanthopterus*, *Ctenochaetus striatus*, *Naso unicornis*, *Caesio caerulaurea*, *Abudefduf sexfasciatus*, *Amblyglyphidodon leucogaster*, *Abudefduf saxatilis*, *Amblyglyphidodon* sp., *Pycnochromis dimidiatus*, *C. ternatensis*, *Dascyllus trimaculatus*, *Pomacentrus albicaudatus*, *Pomacentrus sulfureus*, *Pomacentrus tripunctatus*, *Gomphosus caeruleus*, *Thalassoma purpureum*, *Epibulus insidiator*, *Scarus ferrugineus*, *Scarus niger*, *Chlorurus sordidus*, and *Balistapus undulatus*. Of these fish, half the number of species are not generally classified as plankton feeders. The most unusual sight was to observe parrotfish swimming in loosely formed shoals relatively far away from the reef face in locations more typically occupied by *Caesio* and *Sardinella*. An individual *S. ferrugineus* (rusty parrotfish) was observed to eat a pteropod and then go immediately to graze among dead coral rubble.

These observations are significant for two main reasons. Firstly, it shows that reef fishes' typically recorded feeding (and behavioural) patterns may be considerably modified during certain conditions such as a glut of plankton (and perhaps also a scarcity of food). It is quite likely that such periodic diets provide essential minerals that are generally not so readily available to herbivorous fish species. Secondly, it is clear that zooplankton peaks may suffer intensities of predation that are considerably higher than might be extrapolated from biomass estimations of planktivorous species. For a discussion of vertical currents in the Red Sea, see Yao et al. (2014).

Chapter 19

Suakin Region

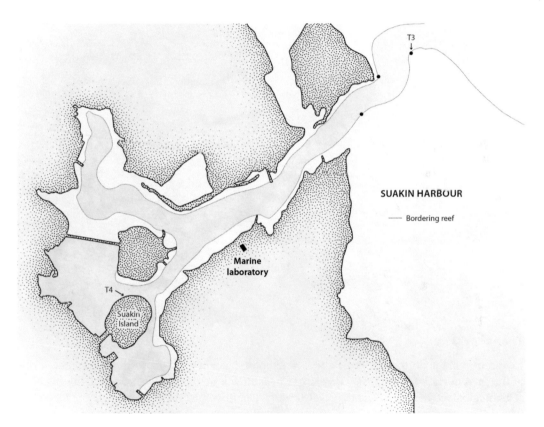

Figure 19.1 Map of Suakin Harbour.
© Vine

DOI: 10.1201/9781003335795-23

Figure 19.2 The ancient city of Suakin was built with coral blocks and mangrove poles.
© Vine

Transect 3

Location: On the northeast side of a reef promontory, near a fixed marker on the southern side of the Suakin Harbour channel. There are two shallow inner basins, one surrounding the main settlement island of Suakin and the other forming a narrow inlet on the north side of Quarantine Island. The latter comprises a long, narrow inlet with bordering reefs extending approximately 3.7 km from Suakin island to the outer beacons. The main channel is narrow and sinuous, varying from about 12–45 m in depth. The northeasterly orientation of its entrance ensures that there is moderate water exchange due to wind-driven currents. This exchange has permitted the development of quite rich coral communities and the presence, even around the island perimeter, of several species more characteristically associated with open-water coral reefs.

The harbour bottom consists of very fine silt, which readily enters suspension during windy periods, rendering the water murky. Reduced light penetration and silting restricted rich coral growth to shallow zones, usually near the reef crest. The lower reef face is generally composed of steep, smooth rock with few live corals. An indication of the gradual reef accretion rate is provided by the old Eastern Telegraph Company undersea cable, which follows the reef edge

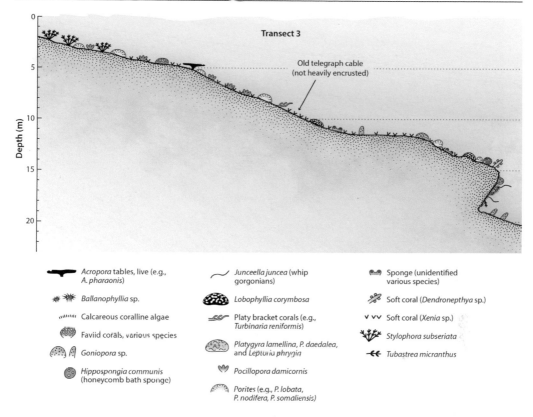

Figure 19.3 Transect 3.

Key (from figure legend):

Acropora tables, live (e.g., A. pharaonis)

Ballanophyllia sp.

Calcareous coralline algae

Faviid corals, various species

Goniopora sp.

Hippospongia communis (honeycomb bath sponge)

Junceella juncea (whip gorgonians)

Lobophyllia corymbosa

Platy bracket corals (e.g., Turbinaria reniformis)

Platygyra lamellina, P. daedalea, and Leptoria phrygia

Pocillopora damicornis

Porites (e.g., P. lobata, P. nodifera, P. somaliensis)

Sponge (unidentified various species)

Soft coral (Dendronepthya sp.)

Soft coral (Xenia sp.)

Stylophora subseriata

Tubastrea micranthus

along the southern side of the channel. The old cable had been submerged for nearly a 100 years at the time of this survey. Yet, probably due to the presence of copper, it was still relatively unencumbered by corals or other invertebrates.

Just outside Suakin Harbour entrance, the fringing reef was vibrant, with a dense cover of live corals together with an abundance of fish life. Large *Plectorhinchus albovittatus* and *Epinephelus tauvina* were common. Many small sharks (*Triaenodon obesus*, *Carcharhinus melanopterus*, and other species) were frequently seen near the reef face or swimming in very shallow water across the top of the reef. In this location, we observed green turtles (*Chelonia mydas*), hawksbill turtles (*Eretmochelys imbricata*), and a solitary dugong (*Dugong dugon*). It was a fascinating place to sit with an aqualung and simply watch the underwater world reveal itself.

Suakin Harbour encompasses a wide variety of biotopes and deserves extensive investigation. The presence, in the 1970s, of Suakin Marine Station (since rebuilt) on the southern side of the harbour channel helped promote such studies. Assisted by the late Dr. J.E. Randall, a list was compiled enumerating fish observed from within the sheltered waters of the harbour.

Morphology: The reef is within the sheltered channel but still receives intense buffeting during storms. The fauna on the gently sloping reef terrace is sparse. The slope levels into a terrace at about 11 m, leading into a gently sloping section from 11 to 15 m and then to a sharply undercut cliff face with a base at 19 m. The seabed consists of very fine silt, easily disturbed and regularly enters suspension during rough weather.

Figure 19.4 Marine turtles play important roles in maintaining balanced and healthy ecosystems, in particular seagrass beds and coral reefs.

© Hans Sjöholm

Figure 19.5 The sea cow or dugong (*Dugong dugon*) in the Red Sea has become a major tourist attraction; this inevitably disturbs these normally shy and highly endangered marine mammals.

© Poelzer Wolfgang/Alamy

Corals: The richest area for coral growth on this transect was from 1 to 3 m, where *Pocillopora*, *Stylophora*, and *Porites* predominated. Corals were sparse below 4 m. Several encrusting species were present on the upper section of the main slope at around 7 m depth. There were isolated outcrops of *Pocillopora* down to 10 m where the old Eastern Telegraph Company cable was still clearly visible. On the steep, undercut, reef face, there were numerous *Junceella* (whip gorgonians) and isolated *Balanophyllia* polyps. Sediment-tolerant species, such as *Lobophyllia*, *Goniopora*, *Favia*, *Porites*, *Platygyra*, and various alcyonarians, predominated below 10 m.

Fish: While herbivorous fish such as *A. sohal*, *C. striatus*, and *S. lividus* were abundant in the shallows, carnivorous species such as *P. diacanthus*, *Plectorhinchus schotaf*, and *P. asfur* were more abundant in deeper water. The butterflyfish *C. larvatus* was characteristic of this biotope, which is somewhat comparable to areas within the inner reaches of Dungonab Bay, where it is almost the only butterflyfish species present. Midwater plankton feeders included *Caesio* sp. and *Naso brevirostris*, while substrate-associated planktivores included *P. squamipinnis*, *P. dimidiatus*, *Dascyllus aruanus*, *Amblyglyphidodon* sp., and *P. tripunctatus*.

Transect 4

Location: Close to the Governor's rest-house wharf on Suakin island.

Morphology: It consists of a smooth, gently sloping muddy bottom with dense beds of *Halimeda* towards the base of the slope. The bottom at 8 m is fairly level and consists of very fine silt or mud with large burrows.

Faunal notes: There were many embedded stones on the slope that, when removed, were found to have live serpulid tubeworms attached to them, despite the fact that the stones were covered with sediment. Sabellid tubeworms are often attached to fragments of pottery embedded in the mud. In this region, several old bottles have been unearthed that date back to the early part of the 20th century, when the island was a busy town, largely constructed from coral rock, and frequented by a cosmopolitan population of traders and British troops. In the early 1970s, the island was almost completely deserted, and all the buildings were in ruins.

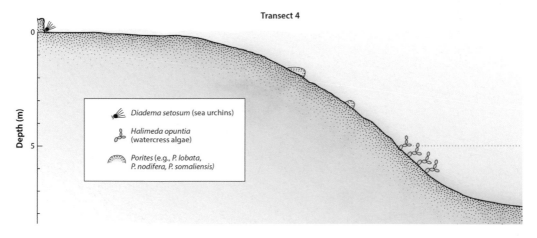

Figure 19.6 Transect 4.

© Vine

Numerous crevices among the boulders forming the harbour wall are inhabited by the black-spined sea urchin, *D. setosum*, which emerges at night to graze on algae. Commensally associated with these sea urchins is the hippolytidae shrimp *Saron neglectus* De Man; its colourful disruptive camouflage makes it appear to be armed with epizoic anemones.

Fish: Fish were not particularly abundant on the deeper section of the transect, whereas there was a remarkable variety of species in the shallows, close to the harbour wall. Mullet and *Caesio* were feeding in shoals. On certain summer occasions, large adult milkfish (*Chanos chanos*) were seen, together with shoals of *Rastrelliger kanagurta*, close to the island.

Chapter 20

Dungonab Bay

The main physical features of Dungonab Bay were first described by the late Dr. Cyril Crossland (1919, 1956) and by William Reed (1964), both of whom lived and worked within the area for prolonged periods of time. Crossland initially recognised it as suitable for collecting pearl oyster spat and growing them into harvestable mother-of-pearl shells. It has several ecological features that make it a unique biotope within the Sudanese Red Sea. Some of these interesting faunistic features of the sheltered inner bay are listed here:

- Corals are not generally well developed within the sheltered bay.
- *Galaxea* is by far the most dominant genus of hard coral in the bay area. This is of interest since *Galaxea* has been identified as a sediment-tolerant species with the potential to play a significant role in reef recovery after damage associated with global warming (Lin et al. 2017).
- Coral reef fish common outside the bay may be absent from, or rare within, the basin.
- There are few species of butterflyfishes in the inner bay; the dominant species is the coral-livorous species *C. larvatus*.
- Despite the scarcity of hard corals in the bay, the coral-predating CoTS (*Acanthaster planci*) was relatively conspicuous there in 1974, feeding on *Xenia* instead of hard coral.
- A 2004 survey done 30 years after the author's original study (1974) indicated that *A. planci* was rare within the bay (Beyer et al. 2015a).
- Although *Pinctada margaritifera* spat was collected near the bay entrance, and growth occurred within the bay, wild pearl oysters were not particularly common.
- The echinoid *D. setosum* was locally distributed, and its feeding activities denuded hard substrata of algae, coral planulae, and many other invertebrate larvae.

These and other relevant aspects of the bay's marine biology are described under Transects 5–13.

Transect 5

Location: Reef slope near Ras Abu Hunter, on the east side of Dungonab Bay.

Morphology: Gently sloping, generally bare, hard substrate incline extending from a shallow terrace at 4 m depth to a deep sandy bottom at 15–17 m. From 8.5 to 9.0 m, the substrate consisted of relatively smooth hard rock with very occasional stones or small isolated coral colonies. At 10 m, a sharp, well-defined undercut ledge (probably formed during the glacial period) provided cover for many invertebrates and fish. A rocky slope from 10 to 17 m was somewhat

DOI: 10.1201/9781003335795-24

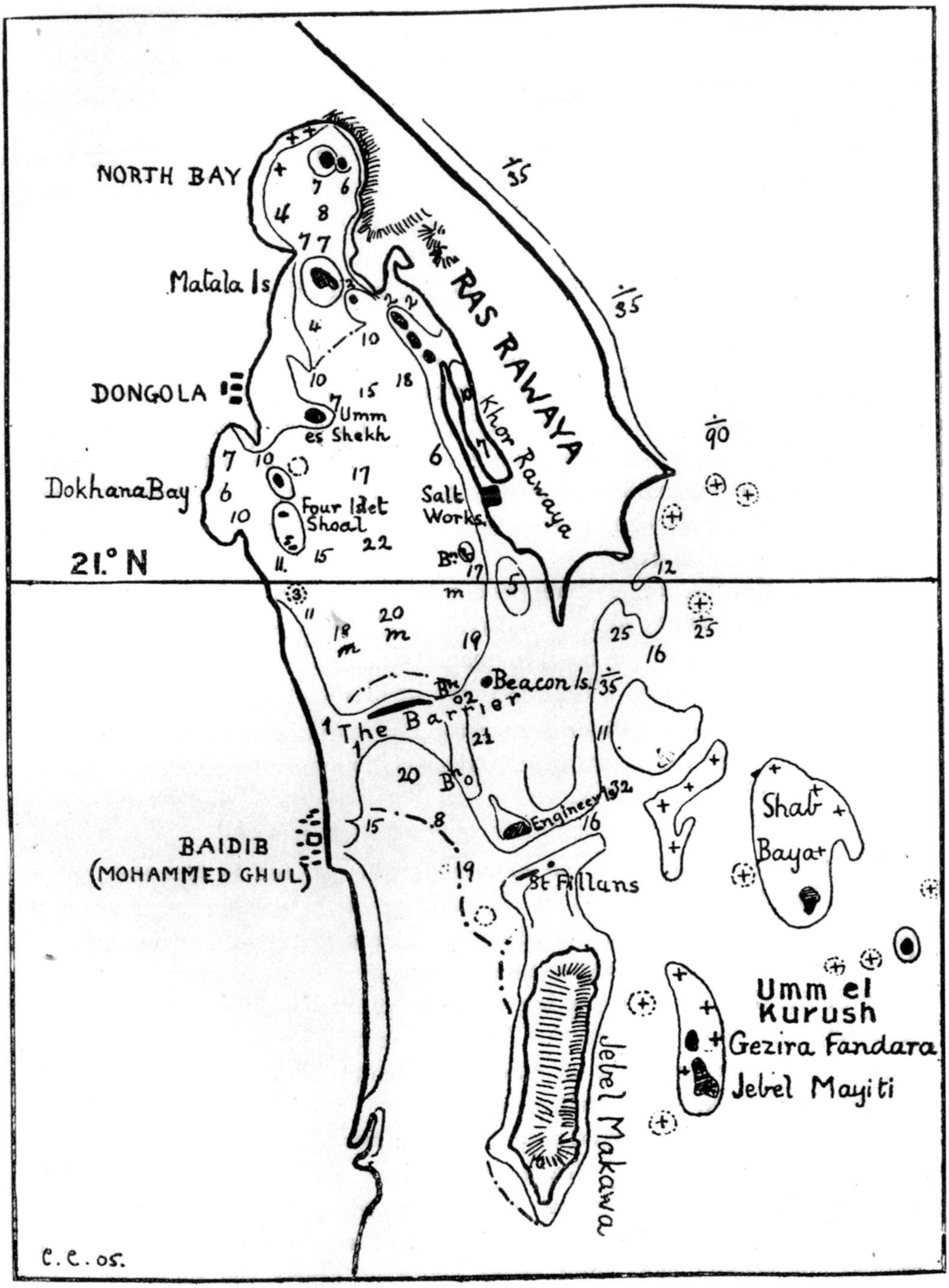

MAP OF DONGOLA AND ITS VICINITY.

Figure 20.1 Map of Sudanese coast and reefs was drawn by the eminent field biologist, Dr. Cyril Crossland, who pioneered the farming of mother-of-pearl in the Red Sea.

© Crossland C (1919)

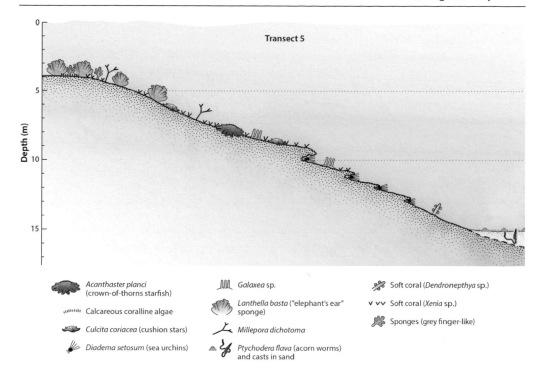

Figure 20.2 Transect 5.

© Vine

steeper and more irregular in outline, with small ledges and crevices. In general form, however, it still lacked cover and had very few fish.

Corals: There was relatively little coral present, and *Galaxea* was the only genus observed (apart from a single isolated bracket of *Acropora* at 4.5 m).

Faunal notes: The shallow flat area at 4 m had a well-defined patch where many upright black finger sponges (tentatively identified as Thorectidae: *Hyrtios erectus*) created a miniature "forest" where fish could find cover from roving predators, such as large *S. barracuda* and *E. tauvina*, which were present here. Red CCA (possibly *Porolithon* sp.) formed thin, loose rubble on the rock surface, and there were many patches of *Xenia macrospiculata*. The cushion star *Culcita coriacea* was noticeable in open situations, extending from the shallow terrace to about 10 m. A large upright, bushy soft coral was host to a pair of anemonefish (*Amphiprion bicinctus*) at 4.5 m.

At 5.0 m, soft sedimentary sand formed a slightly thicker covering on the underlying hard substrate and contained numerous *Ptychodera flava*. This hemichordate (acorn worm) possesses the ability to regenerate its nervous system, including its head. They are abundant in the soft sediments of Dungonab Bay.

At 7.5 m, there was a stand of *M. dichotoma* oriented to face a long-reef current. The seabed lacked any prominent fauna, while a very thin layer of sand covered consolidated beach rock at this depth. In this unexpected locality, we found three specimens of *A. planci*, even though there was no scleractinian coral near the starfish. Unexpectedly, we noted that they were feeding on the soft coral *Xenia*, which was moderately abundant and on which one *A. planci* had its stomach everted. This spiny asteroid is quite common in Dungonab Bay. It has likely adapted its behaviour to take advantage of the abundant growths of *Xenia* and the relative absence of reef-building corals in the semi-enclosed bay. At the time of the survey (1976), this was possibly the first record of *A. planci* feeding exclusively on the soft coral *Xenia* rather than on hard corals.

The upside-down jellyfish (*Casseiopea andromeda*) and *C. coriacea* were the most conspicuous macrofauna present from 8 to 10 m. There were very few fish. At 10 m, a sharp, well-defined undercut ledge provided the best cover (for macrofauna) available on the entire transect. In the shade of the crevice, there were numerous invertebrates and some fish that were not present elsewhere on the transect. Numerous *D. setosum* were gathered in the shade.

At 13 m, a single mother-of-pearl shell (*P. margaritifera*) was found attached by its byssus to the hard seabed. At deeper sections of the reef slope, there were numerous holothurians and casts of *Ptychodera*. The rock slope levelled out between 15 and 17 m, where a soft substrate replaced it with many hemichordate casts and burrows occupied by the goby *Amblyeleotris steinitzi* (together with an alpheid shrimp). The goatfish, *Parupeneus forsskali*, ranged out from the reef and across the muddy bottom.

Fish: Several large predators in the shallow range of the transect included *S. barracuda* and *E. tauvina*. In the absence of any alternative cover, these were utilising the restricted cover presented by a dense area of black sponges. Small fish were notably sparse throughout the transect (probably due to the absence of available hiding places). Groupers, particularly *A. leucogrammicus* and *E. merra*, were quite conspicuous over the whole transect. In this area, these probably depend more on crustaceans than on fish for food. We recorded 19 fish species on the transect, giving an indication of the scarcity of species (cf. 52 at Transect 1 and 60 at Transect 2) relative to the fringing reef near Port Sudan.

Transect 6

Location: Near the ruined building in the lagoon, close to the saltworks tower at Abu Hunter on the southeast side of Dungonab Bay. Set in a shallow lagoonal (back-reef) zone just inshore from the reef top.

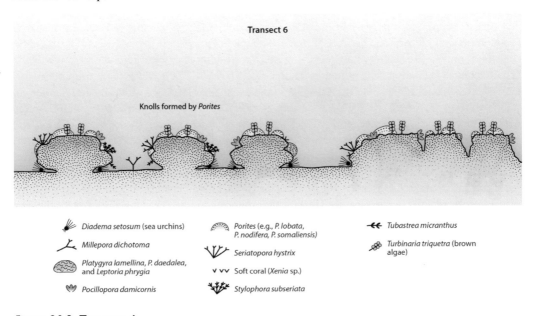

Figure 20.3 **Transect 6.**

© Vine

Morphology: Dominated by many large, flat-topped knolls of *Porites* between which were narrow interconnecting channels.

Corals: *Porites*, *Acropora variabilis*, *S. pistillata*, *Platygyra lamellina*, *Pocillopora damicornis*, and the hydrozoan *M. dichotoma* were the main species. The *Porites* knolls were topped by the brown macroalga *Turbinaria*. The sides of *Porites* knolls had many crevices with colonies of *Pocillopora*, *Xenia*, and, on the more sheltered sides, *Seriatopora*.

Other fauna: The black sea cucumber *Holothuria atra* was frequently observed on coarse white sand between *Porites* knolls. The asteroid *Fromia* sp. was present on algae-encrusted dead *Porites*. The echinoid *D. setosum* was abundant in shaded crevices. Algal grazing by echinoids and fish had left exposed surfaces of coral rubble and dead coral colonies almost bare of algae.

Fish: Twenty-one species were recorded from the area of the transect. This southern area of Dungonab Bay is sufficiently close to the open sea to escape some of the restrictive influences that appear to operate within the bay. On the outer reef slope near the transect area was an assemblage of species more typical of Sudanese coral reefs, including *T. obesus*, *C. fulvoguttatus*, *D. aruanus*, *A. sohal*, and *Chaetodon lineolatus*. None of these were seen in the transect area.

Transect 7a

Location: Approximately 1 km west of the shoreline (in winter) at Abu Salaam, on the eastern side of Dungonab Bay, in an area where the shoreline lacks any distinguishing features.

Morphology: The area appears to be relatively uniform, consisting of an extensive flat zone extending at least 1 km offshore. There appeared (from the surface) to be little holding ground, and we anchored our boat over what seemed to be the rockiest area. The latter turned out to be a well-defined patch of *Xenia* on small pebbles at 7 m depth. This patch rose slightly above the surrounding area of sand. An interesting crevice on this sand is described under Transect 7b.

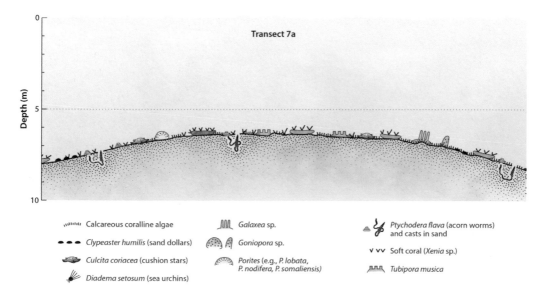

Figure 20.4 Transect 7a.

© Vine

Corals: In the sandy area, there were small colonies of *Tubipora musica*, *Galaxea* sp., *Goniopora* sp., and *Porites*. There were very few hard corals on the *Xenia* patch, but the underside of one stone had a small colony of *Galaxea*.

Faunal notes: The sand had numerous casts of the hemichordate *Ptychodera*. The sand dollar *Clypeaster humilis* was also abundant in the sand. *C. andromeda* was present, but not in large numbers. *C. coriacea* was common and seen feeding on the red CCA, *Porolithon* sp.

Fish: Only 12 species of fish were recorded from the area. Cover was generally lacking except for small crevices among pebbles on the *Xenia* patch. In this situation, there were numerous *P. tripunctatus*. The most notable feature was a shoal of *Paracheilinus octotaenia* in which larger males were displaying by suddenly extending dorsal, caudal, and anal fins while swimming well above the seabed. Simultaneously, smaller females hovered at about 45° to the vertical (with heads pointed upwards) and quivered the posterior edges of their trailing dorsal fins. Approximately one thousand fish were present in the shoal, and more males were apparent than in other shoals observed within the bay. This intensive courtship behaviour coincided with a new moon.

Transect 7b

Location: Approximately 20 m south of the *Xenia* patch described in Transect 7a.

Morphology consisted of a flat slab of rock under which about 30 striped eel catfish (*Plotosus lineatus*) had excavated a cavern.

The tiny catfish were observed to be actively moving stones from their burrows by picking them up in their mouths and carrying them outside the excavation beneath the rock. The two entrances, on opposite sides of the slab, were marked by long lines of rubble that the catfish had shifted. As a result of this burrowing by *Plotosus*, several other animals were able to take advantage of the cover presented in this otherwise bare zone with minimal cover. At one end of the burrow, a small geometric moray eel (*Siderea grisea*) stretched its head out into the open, while immediately next to it was a large hermit crab with anemones attached to its shell. Outside the

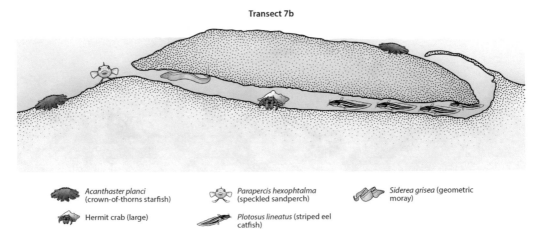

Transect 7b

Acanthaster planci (crown-of-thorns starfish)

Hermit crab (large)

Parapercis hexophtalma (speckled sandperch)

Plotosus lineatus (striped eel catfish)

Siderea grisea (geometric moray)

Figure 20.5 Transect 7b.

© Vine

burrow, a speckled sandperch, *Parapercis hexophthalma*, was sitting on the sand while closely surveying the excavated material for acceptable food items. In a zone where there was almost no suitable cover for animals to hide from predators, *P. lineatus* had created a refuge that was being utilised by a range of other species. CoTS (*A. planci*) was locally abundant, feeding on soft corals.

Transect 8

Location: The south side of the northeast end of Um El Sheikh Island within Dungonab Bay.

Morphology: On a reef platform 2 m deep, a thin layer of white sand covered smooth coral-line rock. There were large flat boulders of dead coral and submerged beach rock with underlying crevices (containing abundant *D. setosum*).

Corals: Throughout Dungonab Bay, corals are relatively sparse, and at this transect, hermatypic species were reduced to very few, mainly an *Acropora* sp. with short upright branches, *S. pistillata*, *P. lamellina*, *Galaxea* sp., and the hydrozoan *Millepora platyphylla*. There were also various soft corals.

Faunal notes: *D. setosum* was common, especially in crevices. It was, in general, a rich area for echinoderms, including *C. coriacea*, *A. planci*, several holothurian species, *Prionocidaris baculosa*, *Echinostrephus molare*, and *C. humilis*.

On the sloping sandy bottom below 3 m were numerous *Ptychodera* casts and alpheid/goby burrows. *C. coriacea* was also present on this sandy seabed, as were several large tridacnid clams.

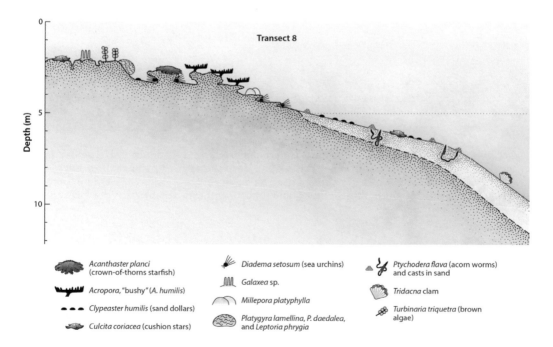

Figure 20.6 Transect 8.

© Vine

Figure 20.7 Two species of giant clams, *Tridacna maxima* and *T. squamosa*, coexist in the Red Sea, with *T. maxima* mostly found near the reef top and shallow edge, while *T. squamosa* occurs in deeper water on the fore-reef slope. They are mixotrophic, capable of both filter-feeding and photosynthesis (thanks to the presence of zooxanthellae).

© Vine

Fish: Twenty-three species of fish were recorded. Larger fish such as the angelfishes *Pomacanthus imperator* and *P. asfur* together with *Heniochus diphreutes* were hiding in shaded holes among boulders. At the same time, pomacentrids swam in isolated, small schools above colonies of *Stylophora*. Here we recorded *Neoglyphidodon melas* juveniles (which until recently had been mistakenly named *Abudefduf melanopus*). Juveniles and adults exhibit quite distinct colour patterns. *P. tripunctatus* is the dominant pomacentrid within Dungonab Bay and was abundant at this transect. Certainly, the most common chaetodon in Dungonab Bay was *C. larvatus*. At sites within Dungonab Bay, it was usually the only butterflyfish recorded, and the lack of flourishing hard corals within the bay is related to the absence of most butterflyfishes. The only other chaetodontidae seen within the major part of the bay were *Chaetodon semilarvatus* and *Heniochus* sp. The most characteristic moray eel in the bay is *S. grisea*, which occurred on this transect in association with *Diadema*.

Transect 9

Location: 1 km west of Um El Sheikh Island in Dungonab Bay.

Morphology: No reefs are visible from the surface in this area. The seabed was uniformly flat and probably representative of a significant area of the bay; it consisted of fine grey mud with many burrows and was around 20 m deep.

Fauna: Occasional large cerianthid anemones were present, but no other fauna was observed.

Transect 10a

Location: Approximately 1 km east of Dungonab village in Dungonab Bay.

Morphology: A shallow submerged patch reef rising from a sloping sandy bottom about 5 m deep to a level reef top at 6 m. Most of the reef top consisted of dead coralline rubble colonised by *Xenia*.

Corals: Very poorly developed compared to reefs outside Dungonab Bay, but more coral diversity at this site than in most other areas visited within the bay. The species recorded were *T. musica*, *M. dichotoma*, *Fungia* sp., *L. corymbosa*, and *Acropora* sp.

Faunal notes: The sand had numerous *Ptychodera* burrows and patches of densely aggregated *D. setosum*. Sea cucumbers, *Holothuria atra*, were scattered over the sand and on the reef. On one side of the reef was a large, prominent knoll (described in Transect 10b) covered by *D. setosum*.

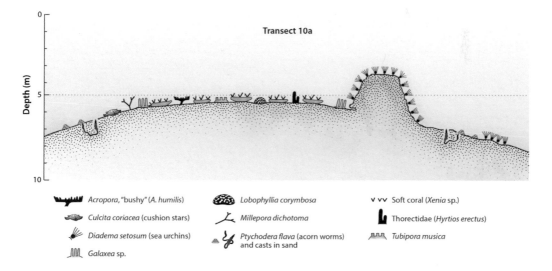

Figure 20.8 Transect 10a.

© Vine

On the reef were many sponges, clams, small echinoids, and large cushion stars (*Culcita*). Recent mass mortalities of cultivated pearl oysters (*P. margaritifera*) within the semi-enclosed bay close to Dungonab village may have affected mother-of-pearl shells in this area since the only specimens that were seen appeared to have been recently killed.

Fish: Restricted assemblage of reef fish (19 species) recorded. The only butterflyfish present were *C. larvatus* and *H. intermedius*. This characteristic reduction in butterflyfish is a feature of almost the entire sheltered bay area. The presence of large predators such as *S. barracuda*, *C. fulvoguttatus*, and serranidae underlines the importance of protective cover for the small damselfish and other species that inhabit the reef. The general paucity of herbivorous reef fishes such as parrotfishes and surgeonfish was probably due in part to overfishing (see Loh et al. 2015) and the very dense aggregations of echinoids, which kept rock surfaces bare of algae and incidentally removed many young invertebrates, thus restricting colonisation by corals.

Transect 10b

Location: A large knoll situated on the southwest side of the patch described in Transect 10a was notable as the habitat of hundreds of black-spined sea urchins, *D. setosum*.

Morphology: The knoll, about 2 m high and about 4 m in diameter, was formed from consolidated dead coral with an extremely worn and pitted surface.

Corals: Only three tiny patches of live *Galaxea* were observed on the knoll, and no other corals were present.

Faunal notes: In almost every gouged hole over the entire knoll, there was a *Diadema* sea urchin. The constant rasping by *Diadema* as they scraped the rock surface for algae had resulted in considerable erosion of the surface of the knoll and the almost complete absence of attached sedentary invertebrates. The *Diadema* were so numerous, and their feeding activity was so intensely concentrated, that they excluded virtually everything else. The knoll's

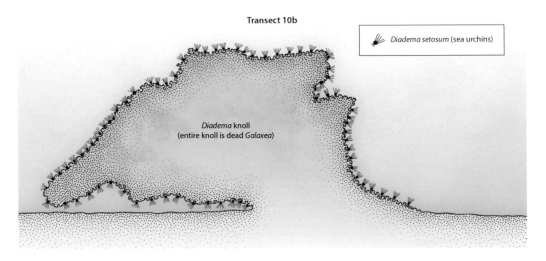

Figure 20.9 Transect 10b.

© Vine

original structure was likely formed by the coral *Galaxea* in a similar way to the formation of the Mesharifa channel knoll described under Transect 25. Activities of *Diadema* had, however, long ago disguised the original form of the knoll.

Fish: Very few fish were present around the knoll, but several species were represented. *C. larvatus* and *H. intermedius* were once again the only recorded butterflyfish.

Transect 11

Location: A low-profile sand cay with sparse vegetation, close to a small island within Dungonab Bay and 100 m out from the northern beach, opposite the midpoint of the island.

Morphology: Below 5 m, the soft bottom consisted of fine white mud with interspersed coarse calcareous fragments. The reef slope was gradual, and the substrate consisted of submerged beach rock thinly covered by sand. Patches of cover were afforded by sponges and soft corals, as discussed in the following.

Corals: There was a dense patch of soft corals from 5 to 6.5 m. This afforded cover for fish. Hard corals were generally scarce, but several faviid colonies were present around 4 m, together with a few colonies of *Stylophora* at 2 m.

Fish: Only seven species were recorded. Even among the soft corals, relatively few species were present, and *E. merra* and *Thalassoma scapularis* were the main species. The *Stylophora* colonies at 2.0 m had *Dascyllus marginatus* associated with them. The absence of pomacentrids from the rest of the transect reflects a lack of suitable cover rather than any shortage of suitable food. Between 7 and 5 m were occasional *P. forsskali*.

Faunal notes: There were numerous casts of *Ptychodera* in the soft sediment below 8 m, together with occasional outcrops of an unidentified green sponge, several *Murex* spp., and an

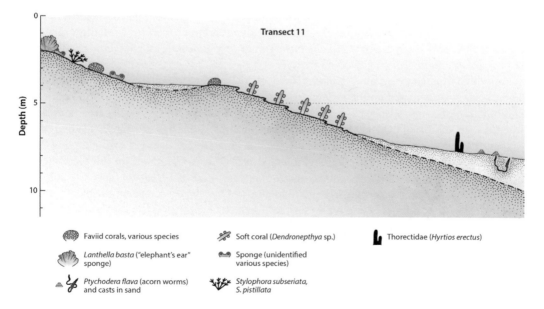

Faviid corals, various species

Soft coral (*Dendronepthya* sp.)

Thorectidae (*Hyrtios erectus*)

Lanthella basta ("elephant's ear" sponge)

Sponge (unidentified various species)

Ptychodera flava (acorn worms) and casts in sand

Stylophora subseriata, S. pistillata

Figure 20.10 Transect 11.

© Vine

Figure 20.11 Osprey chicks camouflaged at the nest.
© Vine

unidentified flatworm. In sandy patches above and below the soft coral patch, there were abundant *Ptychodera* and, at shallow depths, the sand dollar urchin *C. humilis*. In shallow areas less than 2 m, elephant ear sponges (*Ianthella basta*) were very common, and numerous nudibranchs (*Chromodoris quadricolor*) grazed on these.

As with virtually every other island visited off the Sudanese coast, ospreys were present. Their nests were situated on the ground (as is usual in this area) and were constructed from a combination of twigs, sponges, flotsam, and various skeletal remains. The nest areas are usually littered with sun-dried fish, especially triggerfish. Multiple sightings of ospreys feeding off the Sudanese coast indicate that they take many species of coral reef fish. For example, on Harvey Reef among the Towartit reef complex, a single osprey was observed, on three consecutive mornings, to eat *A. sohal*, *C. fasciatus*, and *Rhinecanthus assasi*. Reef-top species such as *A. sohal* and *C. striatus* are particularly prone to attack. Ospreys do not seem to select prey fish species prior to attack, and certain easy-to-catch fish, such as triggerfish, are taken frequently but not digested since their thick skin is unpalatable. This would explain the abundance of whole triggerfish close to osprey nests.

Crabs and Sharks

The north side of the island, exposed to prevailing winds, had a well-developed beach-rock platform and a sandy beach with numerous burrows of the ghost crab *Ocypode saratan*. Tracks of the hermit crab *Coenobita scaevola* led inland to find shade and refuge among the vegetation.

Figure 20.12 Hermit crab, *Coenobita scaevola*, on the beach.
© Vine

Several small blacktip reef sharks were cruising through the shallow water, often only a metre or less from the beach. Dorsal and caudal fins were projecting above the water. We suspect that these sharks were searching for ghost crabs that were observed to enter the sea and remain stationary (e.g., when approached by an osprey or man) rather than always seeking escape in their burrows. Spaet et al. (2012) published a review of elasmobranch research in the Red Sea.

Transect 12

Location: West of Abu Claub within Dungonab Bay. A wide, flat, shallow zone extends off-shore, and coral growth is sparse. The boat was anchored at 11 m over a submerged beach-rock platform with occasional sandy patches.

Morphology: Almost smooth, sloping seabed. The substrate appears to be ancient submerged beach rock and consists of a relatively smooth crust with crevices at certain levels. The reef slope from 5 to 13 m had many undercut ledges, particularly around 10 m.

Corals: Among the hard corals, only *Galaxea* was observed. The soft coral *Xenia* was, however, abundant.

Faunal Notes: The hard substrate restricts benthos to those organisms that can survive either in the very thin patches of sand or wedged in small crevices within the rock. Fanning sand, which was only 1 or 2 cm thick, with one's hand revealed the hemichordate *P. flava*. In addition, several predatory gastropods and the echinoid *C. humilis* were present. Many *D. setosum* were hiding in the daytime in the crevices at 10 m. In addition, feeding tentacles of holothurians

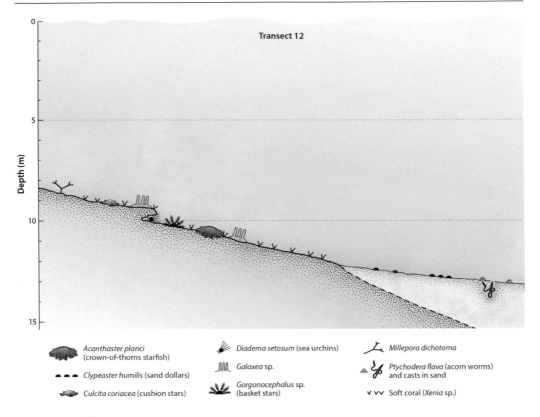

Depth (m)

0

5

10

15

Transect 12

Acanthaster planci
(crown-of-thorns starfish)

Clypeaster humilis (sand dollars)

Culcita coriacea (cushion stars)

Diadema setosum (sea urchins)

Galaxea sp.

Gorgonocephalus sp.
(basket stars)

Millepora dichotoma

Ptychodera flava (acorn worms)
and casts in sand

v vv Soft coral (Xenia sp.)

Figure 20.13 Transect 12.

© Vine

protruded from beneath the ledges, and crinoids and basket stars (*Astroboa nuda*) were also present. Numerous *C. coriacea* were exposed on the reef terrace, and it was somewhat surprising, given the sparseness of hard corals (only *Galaxea* was recorded), to find that the CoTS, *A. planci*, was quite common.

Three specimens of *A. planci* were found at this transect, aggregated under one rock. A fourth specimen was seen in the same proximity, and more were observed at several other sites within the bay. It must rate, alongside *Culcita*, as the most common seastar there. On several occasions, *A. planci* in the bay were found with stomachs everting on *Xenia*. In view of the rapid growth of *Xenia*, and other soft corals, on the skeletons of hard corals that have been recently killed by *A. planci*, this ability to live on the diet of *Xenia* is notable. As long ago as 1919, the marine biologist Cyril Crossland wrote in a semipopular paper entitled *Dangers of Pearl Diving* on the basis of his experiences in Dungonab Bay:

> starfish are extraordinarily rare in the Red Sea, the only fairly common species attempting to make up for this by its extraordinary structure. The disc is broad and carries twelve short rays, the upper surface is covered with spines up to an inch long. The beast is an inconspicuous dull green colour, the spines brownish, and might be touched accidentally. The divers regard it as extremely poisonous.

Crossland was referring to *A. planci* and its status as a fairly common species within Dungonab Bay, which appeared to be similar in1919 to our findings in 1974.

Fish: The small serranid *E. merra* was extremely abundant and was one of the most characteristic fish species in the whole of Dungonab Bay. *T. scapularis* was also very common, but the
general paucity of cover, together with a restricted benthos, led to a limited assemblage of fish
(only 11 species were recorded). For example, none of the normally common plankton-feeding
pomacentridae were present, despite no apparent shortage of nutritious plankton.

At 13 m, there were, as at Transect 7, more *P. octotaenia* females displaying over a *Xenia*-
carpeted seabed. They were hovering about one metre above the seabed in a posture approximately 45° to vertical, while quivering the posterior trailing edges of their dorsal fins. However,
males were not seen on this occasion but were later observed elsewhere in the bay (see Transect
7). Below 13 m, the muddy bottom sloped gradually and contained many hemichordate mounds
together with alpheid/goby (*Amblyeleotris steinitzi*) burrows.

Transect 13

Location: The southeast side of Ras Abu Claub (inside Dungonab Bay), within the localised
shelter of a small, shallow bay.

Morphology: The flat, muddy seabed was approximately 8 m deep.

Corals: None were seen.

Faunal notes: Visibility underwater was affected by extremely fine silt in suspension,
restricting a diver's view to as little as 10 cm. There were many well-formed burrow entrances
approximately 4 cm in diameter. Unfortunately, we did not see the creature responsible for
creating them.

Fish: None were seen.

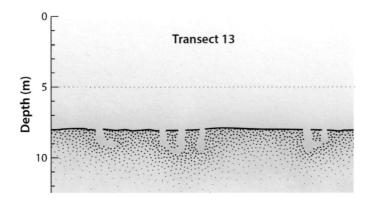

Figure 20.14 Transect 13.

© Vine

Chapter 21

MV. Mani

Transect 14

Location: Starting from the stern section of the wrecked ship, *MV. Mani*, working in a westerly direction towards Port Sudan's large grain elevator.

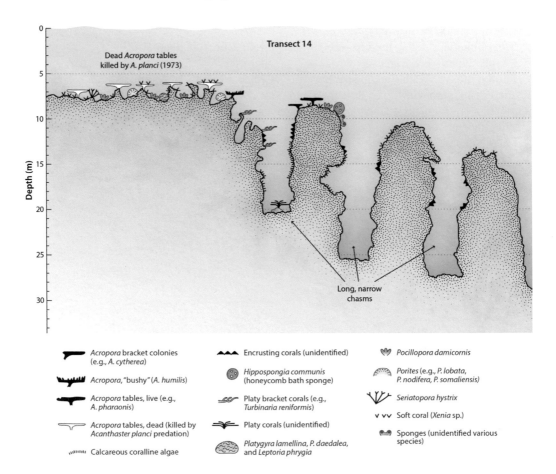

Figure 21.1 Transect 14.

© Vine

DOI: 10.1201/9781003335795-25

Morphology: The transect commences with a gradual slope to 10 m depth. There, abruptly, a series of interconnected steep-sided deep holes or gullies began with a general direction parallel with the line of the reef crest rather than perpendicular to it (which is more usual with erosional or growth features such as these). The base of the first gully was at 21 m and consisted of very fine silt, some of which was in suspension. A narrow ridge separated the first gully from the second gully, with a base at 28 m. This pattern of holes extended over a wide area.

Corals: The reef next to the stern of the *Mani* had considerable debris. A number of large *Acropora* tables, killed by *A. planci* between 1972 and 1974, were still intact structurally. The surface of the terrace was very irregular, with numerous small pits. There was a great deal of *Xenia* covering dead coral on the terrace, in addition to live colonies of *Pocillopora* sp., brackets of *A. cytherea*, *Seriatopora*, *Porites*, and *P. lamellina*. There were relatively few coral colonies in the gullies, although several platy species encrusted the upper walls. On the ridge between the first and second gullies, live tables of *Acropora* had escaped *A. planci* predation, possibly due to the intervening gully. *Xenia* was also abundant here.

Faunal notes: *A. planci* was not seen during this particular survey but was known to have been active in the area from 1972 to 1974. Several sponges were present on the transect, particularly on the vertical gully walls.

Fish: Most of the recorded fish species inhabited the shallows. There were few fish in the chasms. Only *Chlorurus gibbus*, *C. striatus*, and *C. fasciatus* were seen halfway down the gully, with no fish recorded from around the base level.

Chapter 22

Towartit Reefs

Transect 15

Location: On the northwest side of a shallow patch reef within the area protected by North Towartit beacon reef.

Morphology: A number of large knolls give way to a long, more or less level, sandy terrace on the north side of the reef at about 11 m depth. The terrace is formed from smooth coral rock with sandy patches.

Corals: Several large tables of *Acropora* were covered by fluffy masses of a fine, loosely attached green alga (cf. *Hydroclathrus*), which was apparently smothering some corals at the time of our investigations (15 November 1975). Colonies of *Pocillopora* were inhabited by schools of *D. aruanus*. Much of the terrace was dominated by *Xenia*. Colonies of *Porites* and *Goniastrea* were quite common. See Vermeij et al. (2011) for notes on the impact of CCA and *Ulva* on macroalgae in Hawaii. They demonstrate that CCA can suppress macroalgal growth to the benefit of reef-building corals.

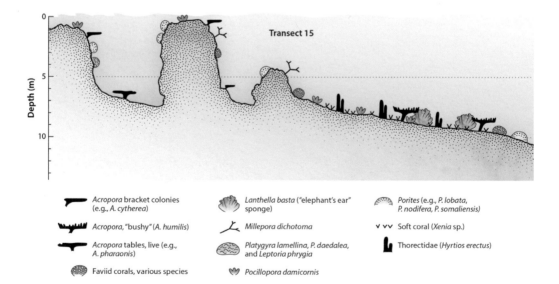

Figure 22.1 Transect 15.

© Vine

DOI: 10.1201/9781003335795-26

Faunal notes: Several sponges were common, including a black finger sponge (tentatively identified as Thorectidae: *H. erectus*) and an elephant's ear sponge. The general reefscape was one of barrenness.

Transect 16

Location: The southeast side of Red Beacon reef in the Towartit reef complex.

Morphology: A number of coral knolls close to the reef are followed if one swims in a south-easterly direction by a long, gently sloping, stepped terrace ranging from 10 m to more than 20 m depth over a horizontal distance of about 200 m.

Corals: *Xenia* is the most abundant soft coral. *Pocillopora* and *Porites* are the dominant hard corals present.

Faunal notes: The terrace is very poor in terms of variety of marine life, as it is dominated by *Xenia*.

Fish: Fish distributed around the shallow-water knoll at the upper end of the transect are detailed in Transect 17.

Transect 17

Location: On Transect 16.

Morphology: Consists of a large knoll formed mainly from *Porites*, with *Pocillopora* dominant on the flat, wave-cut reef top.

Fish: The majority of fish were concentrated on the side of the knoll facing into the wind-driven surface current. *C. ternatensis*, *C. caerulea*, and *P. dimidiatus* formed the main shoal in shallow water. *C. ternatensis* usually extended farthest from the protective cover of the knoll,

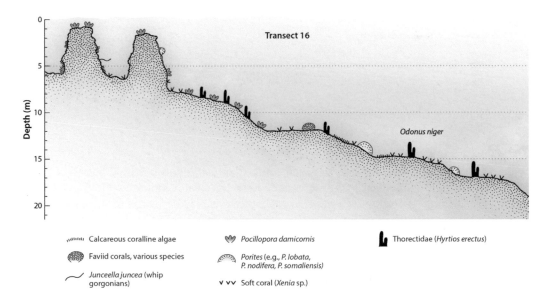

Figure 22.2 Transect 16.

© Vine

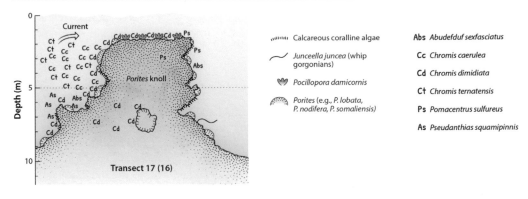

Figure 22.3 Transect 17.

© Vine

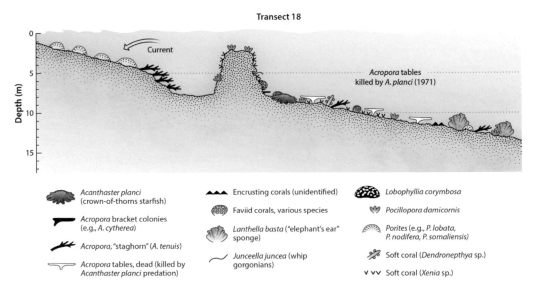

Figure 22.4 Transect 18.

© Vine

while *P. dimidiatus* always kept close to the coral. *P. squamipinnis* occurred below the main school of the three damselfishes, with some individual jewelfish venturing up to 2 m from the coral while feeding on plankton. *P. sulfureus* is another conspicuous plankton feeder, but it was more closely associated with the substrate than any of the four species discussed earlier. Shoals of pomacentrids and *Pseudanthias* form an effective feeding net, and those farthest from the coral can catch larger plankters, while those nearer it are left to feed on smaller plankters. Competition for food is both intraspecific and interspecific, thus providing *C. ternatensis* (some of which are usually on the outer extremities of the feeding school) with a potential advantage. This advantage is offset, however, by the increased likelihood of predation among those species that venture farthest from refuge. Occasional personal observations on the food of small jacks (carangids) have shown that *C. ternatensis* is frequently eaten by these predators.

Transect 18

Location: On Harvey Reef (on which the CCSRG platform stood) inside the Towartit reef complex, at approximately the midpoint of the easterly side of the reef.

Morphology: The shallow reef face is rather exposed to northerly winds, which cause steep short waves despite the protection of the outer barrier reef. The reef face consists of a relatively gradual upper section of reef slope, with *Porites* coral heads and a thicket of bushy *Acropora* sp. towards the base of the slope. There is a sand and rubble zone at 8 m, and this continues along most of this region into a gently sloping terrace with several knolls arising from 8 m to within about 1 m of the surface (at winter tide levels).

Corals: As stated earlier, *Porites* and *Acropora* are common in the shallows. The large knolls are also formed mainly of *Porites*, and one knoll is described in Transect 19. Live corals were rather poorly represented on the terrace, although small colonies of *Porites* were quite numerous.

Figure 22.5 The Titan triggerfish, *Balistoides viridescens*, is an opportunistic feeder on a variety of echinoderms and other invertebrates, seen here feeding on crown-of-thorns, *A. planci*.

© Vine

Other fauna: *A. planci* was observed at this site, and there was an aggregation of these starfish present in the area from 1971 to 1973 (see discussion here).

Fish: *A. sohal*, *Callicanthus lituratus*, *Z. veliferum*, and several species of parrotfishes were present here. A school of 40 or more large humphead parrotfish (*Bolbometopon muricatum*) was resident on the reef and was seen feeding in very shallow water among the *Porites* knolls. On such occasions, their caudal and pectoral fins often broke the surface, giving a false impression, from a distance, of sharks.

Near the shallow reef base and in upper areas of the terrace, two large triggerfish, *Pseudobalistes flavimarginatus* and *Balistoides viridescens*, are relatively abundant. Both of these fish have been observed in this area to attack and eat coral-predating *A. planci*.

Another predator that occurred here was the pufferfish *Arothron hispidus*. Signs of the destruction caused by a feeding aggregation of *A. planci* (present in 1972) were observed on the terrace below the large knoll, where most of the *Acropora* tables were dead and covered by algae. On this deeper section of terrace, there were few fish present, and no large individuals were seen.

Transect 19

Location: The large knoll situated on Transect 18.

Morphology: The knoll is composed primarily of *Porites*. It is flat-topped, with a steep exposed side and a more gradually sloping sheltered side. There is a large cavern on the western side of the knoll, whose inner walls are covered with whip gorgonians and other sciophilous species.

Corals: The main structure of the knoll is *Porites*. There are small colonies of *Pocillopora* on top, together with fans of *M. dichotoma*. The upper wave-exposed slope has bushy *Acropora*, *Xenia*, and *T. musica*. The sheltered face has much bare rock with *Xenia* and a few colonies of *P. lamellina*. *Galaxea* and several other scleractinian corals are present on the inner walls of the cavern.

Other fauna: There were small patches of a red encrusting sponge on the roof of the cavern.

Fish: The behaviour of Pomacentridae and *Pseudanthias* associated with the knoll is the same as that described in Transect 17. A single pair of *C. semilarvatus* was observed to occupy the small cavern (to the right of the main one) for at least three years (1972–1976).

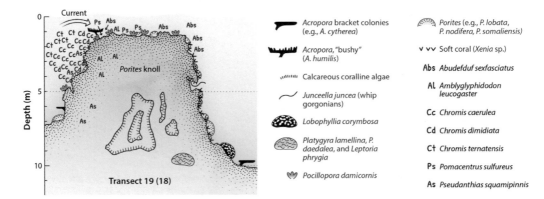

Figure 22.6 Transect 19.
© Vine

Transect 20

Location: On the south end of Little Harvey Reef: that is, a small patch reef immediately south of Harvey Reef (on which the CCSRG platform stood) inside the Towartit reef complex.

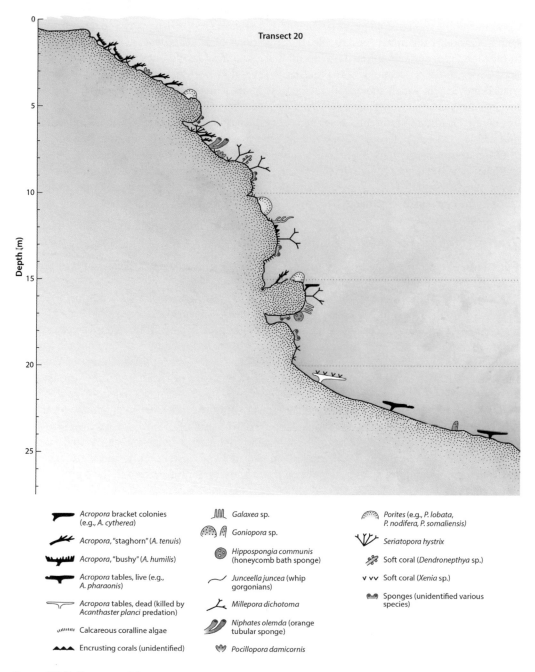

Transect 20

Figure 22.7 Transect 20.

© Vine

Morphology: The shallow zone, from 1 to 5 m, is densely covered with bushy *Acropora*. At 5 m, a large promontory, formed by the *Porites*, juts out, forming an overhang beneath. The steep reef face drops to a crevice and a very narrow terrace at 8 m. The steepest section commences at 8–9 m and extends down to 15 m.

Corals: The shallow zone was dominated by bushy *Acropora* with *Pocillopora* and *Herpolitha*. Coralline rubble was covered by *Xenia*. Staghorn *Acropora* grew on a narrow 8-m terrace, together with *Pocillopora* and much coral rubble. Large colonies of *M. dichotoma* at 8–9 m extended fans to face long-reef currents. From 11 to 15 m, there were few hard corals. *Xenia* was abundant on dead corals in deeper sections of the reef face. Several *Porites* knolls near the reef base created shaded crevices in which sponges flourished. At 16 m, a stand of *M. dichotoma* faced outwards towards the sand, indicating a predominant current perpendicular to the plane of the reef face at this depth. Dead *Acropora* tables were present at the reef base, along with coarse sand and coralline rubble. At 13 m, outcrops of rock with *Acropora* tables and other corals were present.

Other fauna: Many hanging sponges at about 7 m, under the *Porites* knoll.

Fish: In a sheltered crevice at about 8 m, there was a solitary *Pseudochromis flavivertex* (more characteristic of lagoons and harbours), and *Pseudochromis fridmani* was swimming close to the rock on all overhangs.

Transect 21

Location: The northeast side of Laurie reef, one of the Towartit patch reefs south of the reef bearing the CCSRG platform.

Morphology: This is one of the most scenic and interesting reefs we visited in the Sudanese Red Sea. Along the northeast end of the reef, there is a narrow terrace from 10 to 11 m on which there are numerous corals, including large tables of *Acropora*, followed by a very sharply undercut reef face extending from about 11–24 m. This is followed by a second terrace from 24 to 26 m and a second steep reef face from 26 to 40 m. Below 42 m, the sediment consisted of fine white silt with small burrows. There were large outcrops of rock on which various corals were attached.

Further to the south, along the reef, the terrace at 15 m was much broader, and the prominent drop-off was replaced by a more gradual slope from 15 to 23 m. This was followed by a narrow ledge and a further steep drop to 30 m, leading to a narrow ledge and an overhanging reef face from 33 to 38 m. Below 38 m, the seabed sloped away fairly steeply, and there were once again huge rocky outcrops. A large knoll on the shallowest terrace, rising to the surface from 9 m depth, is described under Transect 22.

Corals: The first underhang from 11 to 24 m had a rich assemblage of invertebrates, notably large black coral trees and various gorgonians. Overhanging the crest of the second drop-off were several large *Acropora* tables. Near the base of the reef face at 40 m were many large colonies of *Goniopora* sp. and a rich assemblage of live corals. The water was unusually clear, and there was more brown macroalgae in deep water than we had recorded elsewhere.

Other fauna: The shallowest drop-off was patterned by small, brilliant yellow sponges and numerous ascidians. Both overhangs had rich invertebrate fauna.

Fish: Both *C. striatus* and *C. austriacus* were common as deep as 45 m, and this presence coincided with prolific algal growth at this depth. Swimming among the large boulders on the deepest terrace investigated were schools of the snapper *Lutjanus argentimaculatus* and several deep-water groupers, *Variola louti*. In general, fish were confined to the terraces, and there were very few fish on the steep reef faces.

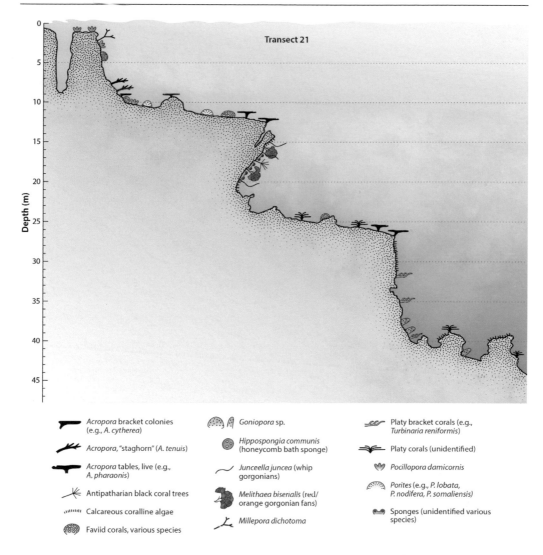

Transect 21

Depth (m)

Acropora bracket colonies (e.g., *A. cytherea*)

Acropora, "staghorn" (*A. tenuis*)

Acropora tables, live (e.g., *A. pharaonis*)

Antipatharian black coral trees

Calcareous coralline algae

Faviid corals, various species

Goniopora sp.

Hippospongia communis (honeycomb bath sponge)

Junceella juncea (whip gorgonians)

Melithaea biserialis (red/orange gorgonian fans)

Millepora dichotoma

Platy bracket corals (e.g., *Turbinaria reniformis*)

Platy corals (unidentified)

Pocillopora damicornis

Porites (e.g., *P. lobata*, *P. nodifera*, *P. somaliensis*)

Sponges (unidentified various species)

Figure 22.8 Transect 21.

© Vine

Transect 22

Location: On Transect 21.

   Morphology: Consists of a north–south section through a large coral knoll. The basic structure of the knoll is formed by *Porites* sp., but other corals have grown among crevices and in situations sheltered from the prevailing wind and waves.

   Corals: The most characteristic feature of the knoll may be related to the influence of prevailing winds. The steep north face was mostly covered by CCA, and there were several small stands of *M. dichotoma* and *M. platyphylla* associated with live *Porites* in the upper two metres. Deeper down on the north face, colonies of the solitary coral *Balanophyllia* sp. were attached to

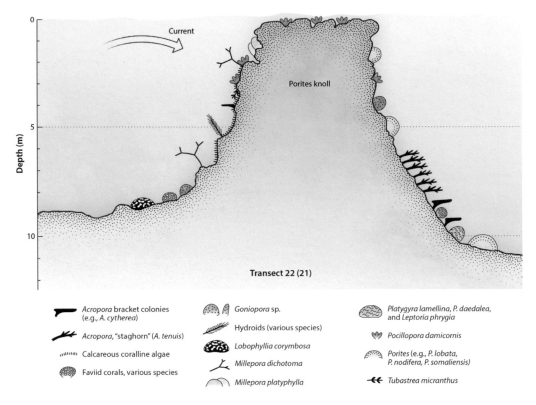

Current

Porites knoll

Depth (m)

0

5

10

Transect 22 (21)

►▬ *Acropora* bracket colonies (e.g., *A. cytherea*)	*Goniopora* sp.	*Platygyra lamellina, P. daedalea,* and *Leptoria phrygia*
➤ *Acropora*, "staghorn" (*A. tenuis*)	Hydroids (various species)	*Pocillopora damicornis*
⋯⋯ Calcareous coralline algae	*Lobophyllia corymbosa*	*Porites* (e.g., *P. lobata, P. nodifera, P. somaliensis*)
Faviid corals, various species	*Millepora dichotoma*	◄◄ *Tubastrea micranthus*
	Millepora platyphylla	

Figure 22.9 Transect 22.

© Vine

roofs of shaded crevices, and the hydroid *Gymnangium eximium* was common on rock surfaces that were well-lit and exposed to relatively strong water movement. Prolific coral growth was absent on the north side. Still, a large stand of *M. dichotoma* was present at 6 m, and several faviidae and other corals were present at the base of the face on a gently sloping reef section leading down to a sandy bottom. The knoll is exposed to prevailing wind and waves from the north, which break over the level reef top, thus creating somewhat sheltered conditions on the southern side. The top of the knoll consists mainly of *Porites* sp. and small colonies of *P. damicornis*. Conditions on the leeward side are more suitable for the settlement of larvae and the growth of less robust corals.

Porites colonies extend deeper down the slope than on the north side and are followed by a sloping reef of branching *Acropora*. Below this are several large brackets of *A. cytherea* and a variety of other corals.

Fish: Fish associated with the knoll were mainly planktivorous. On top of the reef, water movement was so vigorous that only *Thalassoma rueppellii* was present. There were very few fish on the sheltered southern side, where coral growth was most luxuriant. Eight planktivorous fish species formed a multicoloured shoal on the windward side, where they all faced into the current. Such dense shoals act like feeding nets (see discussion under Transect 17).

In Transects 17 and 19, the relative positions of various planktivorous fish are indicated. It has been noted, for example, that shoals of *C. ternatensis* usually extend farthest out and that this is the most common damselfish taken by small jacks, which hunt around the reef edge. *P. sulfureus* (bright yellow) does not move far from cover, and *P. dimidiatus* (vertically striped) extends only slightly further out, always keeping close to the coral. *A. leucogaster* (not shoaling and of a larger size) also tends to keep close, occasionally darting forth to catch a particular plankter. In contrast, *A. sexfasciatus* (shoaling, vertical stripes) may swim out several metres from the coral. *Caesio* spp. are frequently seen feeding in shoals up to 20 m from the coral, and they do not seek refuge in coral crevices but rather tend to hug the reef more closely at the approach of carangids or a barracuda. *Sardinella* sometimes remains close to the reef. On other occasions, when the school is sufficiently large and densely packed with fish, they may be found at considerable distances from reefs, depending on the degree of protection afforded by their schooling behaviour.

Chapter 23

Mesharifa Island

Transect 23

Location: Approximately 5 km northeast of Mesharifa Island are two navigational beacons marking the north and south sides of a channel that leads from the coastal passage, inshore of Mesharifa Island, to beyond the barrier reef. The transect was on the leeward-facing reef slope, close to the north beacon.

Morphology: The reef slope here is inside the protection of the outer barrier reef (which is several kilometres further offshore). It consists of a steep, *Porites*-dominated shallow fore-reef, levelling out at about 10 m, with large knolls rising to 5 or 6 m below the surface, followed by a gradual slope into deeper water.

Corals: The shallow zone was dominated by *Porites*. Fans of *M. dichotoma* were oriented to face a north–south current except at the reef crest, where they met the direction of the surge.

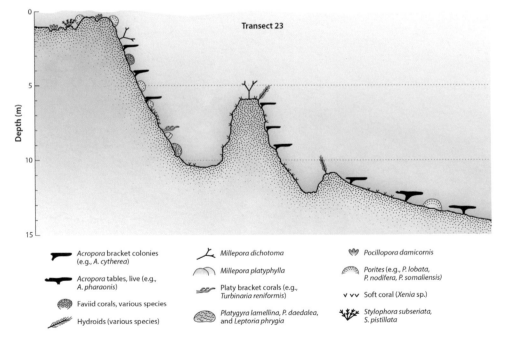

Figure 23.1 Transect 23.

© Vine

DOI: 10.1201/9781003335795-27

Many of the *Acropora* tables occurring at around 13 m depth had markedly convex surfaces. Corals in the shallow zone, down to 13 m, were varied and interspersed with patches of *Xenia*.

Other fauna: *Lytocarpus* sp. and other hydroids, including *G. eximium*, were abundant on the upper surfaces of coral rocks.

Fish: It is an extremely rich and flourishing biotope, supporting large schools of predatory fishes such as *C. fulvoguttatus*, *Caranx melampygus*, and many planktivorous species such as *Sardinella* sp., *P. squamipinnis*, *Caesio* spp., and six species of pomacentridae. This reef section is one of the most likely locations in the Sudanese Red Sea to observe the sizable planktivorous reef manta *M. alfredi*, and this was confirmed on both days when we dived there.

A north-to-south current of approximately one knot carried plankton past the reef promontory, and manta rays preferred this situation for feeding. This is one of the few localities where we observed *C. semilarvatus* grouped in small schools rather than in pairs. It is a planktivorous butterflyfish, and conditions at this site were highly favourable for its feeding. The schools tend to form and break up with the addition and departure of separate pairs.

Transect 24

Location: Mesharifa channel, close to the east end of the island.

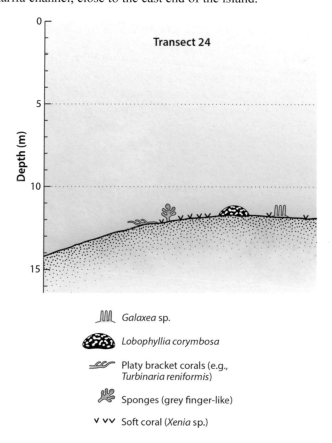

Figure 23.2 Transect 24.
© Vine

Morphology: A fine, whitish silt covers a level seabed at 12 m, and particulate suspension reduces visibility to approximately 5 m on very calm days. During rough weather, visibility close to the seabed may be less than 50 cm.

Corals: It is a characteristically sheltered biotope comparable to the base of the reefs inside Suakin Harbour. *Xenia* is abundant, and the dominant scleractinians are *Lobophyllia corymbosa*, *Galaxea* sp., and *Turbinaria* sp., but coral growth is not well developed.

Fish: Only 15 species were recorded from this somewhat restricted biotope – the small number of herbivorous fish resulting from poor algal growth due to light diminution caused by turbidity. The major fish species present are planktivorous forms that feed on plankton carried by currents that flow through the channel. Several giant manta rays were seen feeding at the surface of this channel on a calm day in early December (1975). *P. flavivertex* and *P. asfur* are particularly characteristic inhabitants of such murky water and sheltered environments. Both were abundant at this transect. *C. striatus* and *H. harid* are the only two herbivores that we recorded. Both are also opportunists and were observed elsewhere feeding on plankton and sessile fauna (see description of Transect 2).

Transect 25

Location: Approximately 100 m southeast of Mesharifa Island.

Morphology: It was dominated by a massive knoll formed entirely by *Galaxea*. The knoll was 5.5 m high and had a diameter of approximately 8 m. Below the knoll, a sandy seabed shelved gradually.

Corals: Such knoll formation by *Galaxea* is not common in the central region of the Sudanese Red Sea, closer to Port Sudan and Suakin, but was frequently seen within Dungonab Bay, where *Galaxea* is the dominant genus of hard coral. Most of the *Galaxea* on the surface of the

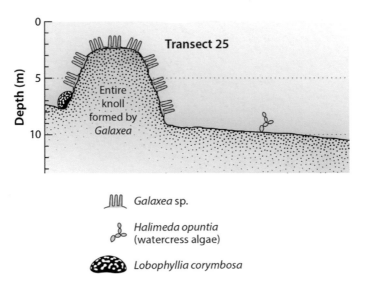

Figure 23.3 Transect 25.

© Vine

knoll was alive, but numerous small patches of dead coral and crevices had formed among the rather loose scaffold structure.

Other fauna and flora: Burrows of the hemichordate *P. flava* were common on the sandy terrace. Alpheid burrows were also present. Numerous small "upside down jellyfish" (*Casseiopia andromeda*) were lying on the sand. A loosely attached, patchy algal covering comprised *Caulerpa*, *Halimeda macroloba*, and *Hydroclathrus*.

Fish: Twenty-one species were recorded, and most of these occurred around the *Galaxea* knoll. Only two species were seen on the sand: an unidentified goby and a sand smelt (*Malacanthus latovittatus*), which darted headfirst into the sand upon being approached.

Chapter 24

Shambaya Island

Transect 26

This transect was situated on the barrier reef in the northern section of the Sudanese Red Sea off a narrow sandspit marked on the British Admiralty charts as Shambaya Island. The island is east of Mayetib Island and is about 150 m long by 30 m wide. There was no visible vegetation, and the only macrofauna comprised hermit crabs, a few ghost crabs (*O. saratan*), and several seabirds. On the leeward side of the sandspit, there was a shallow lagoon in which we observed a shark, about 2 m long, swimming over a very shallow bottom. We also saw in the same lagoon a green turtle grazing on seagrasses. There is a broad, shallow lagoon on the seaward, exposed side, extending about 250 m from the sandspit to the reef crest.

Location: The transect commenced in the mid-portion of the seaward side of the sandspit, across the lagoon, and down the reef face.

Morphology: The shallow area shelved very gradually from the sandspit down to a depth of about 1 m in the back-reef zone. The platform and crest were very turbulent areas. The shallow reef face was deeply incised to form spurs and grooves. At 15 m, the slope became more gradual, and the reef extended into a broad, gently sloping terrace. At 30 m, the slope once again steepened. In this section, and down to 40 m, there were occasional sandy patches located in the undulations of the reef. The base of the main reef was at 40 m, and from there a shelving sandy seabed extended into deeper water. Prominent rubble mounds from about 45–5 m are discussed under the fish section here.

Corals: In the back-reef zone, *Stylophora* and *Seriatopora* were abundant. On shallow coral heads behind the reef top, organ pipe coral (*Tubipora musica*) was conspicuous. Coral growth and fish biodiversity were greater on the spurs than in the grooves (where pink CCA was particularly abundant). On the terrace below 15 m, there were rich coral communities, including many large *Acropora* tables. At 30 m, *Xenia* became dominant and was attached to dead corals and coral rubble. There were also many soft corals and gorgonians on the deep reef face.

The slope from 43 to beyond 50 m was notable for the presence of large, upright, bushy black corals of a species that we had not seen elsewhere.

Fish: *Dascyllus aruanus* was common among branches of *Stylophora* and *Seriatopota* in the back reef. *A. sohal* and *A. sexfasciatus* seem best able to cope with rapid water movement on the outer platform and near the reef crest. They were dominant in these zones. *E. tauvina* was common on sandy patches between 30 and 40 m. *Odonus niger*, the plankton-feeding triggerfish, formed schools between the reef base and sandy seabed.

DOI: 10.1201/9781003335795-28

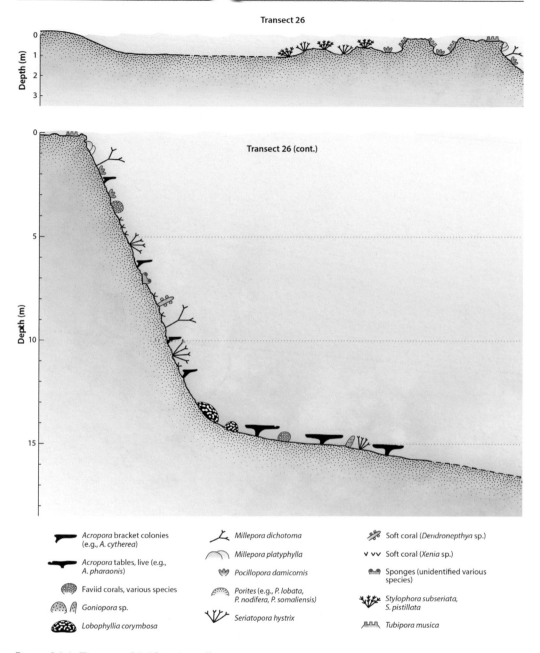

Transect 26

Transect 26 (cont.)

Acropora bracket colonies (e.g., *A. cytherea*)	*Millepora dichotoma*	Soft coral (*Dendronepthya* sp.)
Acropora tables, live (e.g., *A. pharaonis*)	*Millepora platyphylla*	∨ ∨∨ Soft coral (*Xenia* sp.)
Faviid corals, various species	*Pocillopora damicornis*	Sponges (unidentified various species)
Goniopora sp.	*Porites* (e.g., *P. lobata, P. nodifera, P. somaliensis*)	*Stylophora subseriata, S. pistillata*
Lobophyllia corymbosa	*Seriatopora hystrix*	*Tubipora musica*

Figure 24.1 Transect 26 (Continued).

© Vine

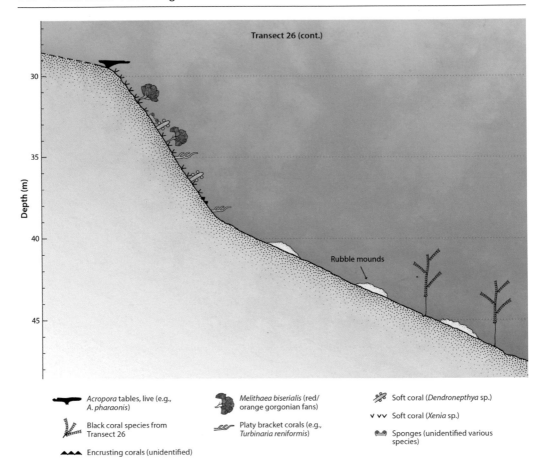

Transect 26 (cont.)

![Acropora symbol]	Acropora tables, live (e.g., A. pharaonis)
![Black coral symbol]	Black coral species from Transect 26
![Encrusting corals symbol]	Encrusting corals (unidentified)
![Melithaea symbol]	Melithaea biserialis (red/ orange gorgonian fans)
![Platy bracket corals symbol]	Platy bracket corals (e.g., Turbinaria reniformis)
![Soft coral Dendronepthya symbol]	Soft coral (Dendronepthya sp.)
v vv	Soft coral (Xenia sp.)
![Sponges symbol]	Sponges (unidentified various species)

Figure 24.1 (Continued)

The deepest sandy zone had many mounds of coralline rubble, each about 30 cm high. Above each swam the blue sand tilefish, *M. latovittatus*. These retreated into discrete holes in the rubble mounds when we approached. They feed by squirting water at the sand, revealing buried worms and other creatures. They live in open sand areas with patches of low-profile reef and rubble at depths of 5–70 m (usually less than 15 m) and are often seen as monogamous pairs that maintain a burrow. Clark et al. (1998) described the large foraging areas of pairs (over 900 sq m) and studied their reproduction.

Chapter 25

Wingate Barrier Reef

Transect 27

Location: Situated at the most southerly point of Wingate reef near Port Sudan.

Morphology: The reef has a steep, vertical, and, in places, overhanging face that descends from just below the crest to 20 m. There is a fairly narrow sandy area between the reef base at 20 m and a raised terrace, which begins at 17 m.

Corals: The reef face is steep or undercut, with few hermatypic corals growing on it. It is, however, rich in sponges, gorgonians, and soft corals. The crest is formed mainly by *Porites*, and in the upper 4 m there are several corals, including *Pocillopora*, *Goniastrea*, *Acropora*, *Porites*, and *Lobophyllia*. On the underhanging reef face, especially from 4 to 8 m depth, there are tube sponges, and the ahermatypic coral *Tubastrea micrantha* is abundant. *Xenia* is present on much of the slope, especially from 11 to 15 m. Orange gorgonians and whip corals cover the shaded roofs of a deep, in-cut horizontal crevice at 11 m.

The 17-m terrace has a great deal of *Xenia*, with small coral colonies interspersed, including bushy *Acropora*, various faviidae, *Porites*, and *Pachyseris*. On the outer slope of the terrace, at 18–20 m, there are colonies of *M. dichotoma*.

Fish: Most of the fish were concentrated on the terrace and near the seabed, but certain species were characteristic of the steep or undercut reef face, notably *P. squamipinnis*, which existed in large schools, from the reef base to within 4 m of the surface. *P. fridmani* was very common in all under-hanging sections, and *Cephalopholis miniata* was also common and seemed to be actively hunting *Pseudanthias*. In addition, there were other fish on the reef face, for example, *C. ternatensis*, *P. dimidiatus*, *T. lunare*, and *Paracirrhites forsteri*. Nearer the reef crest, in the upper 4 m, *P. dimidiatus* replaced *Pseudanthias* as the dominant plankton feeder, and *T. rueppellii* was common. On top of the reef, *A. sohal*, *T. rueppellii*, *P. tripunctatus*, and *C. striatus* were the dominant species.

Transect 28

Location: Situated on the east side of the Wingate barrier reef, approximately 1 km southeast of the North Wingate beacon.

Morphology: The reef top is relatively exposed, although it does not receive the full onslaught of the northerly winds that impact the northwest side of the reef. The reef face is deeply incised with spurs and grooves, and there are many knolls separated from the main reef. The shallow face is rather undercut. The main stepped terrace slopes from 5 to 18 m and gives way to a

DOI: 10.1201/9781003335795-29

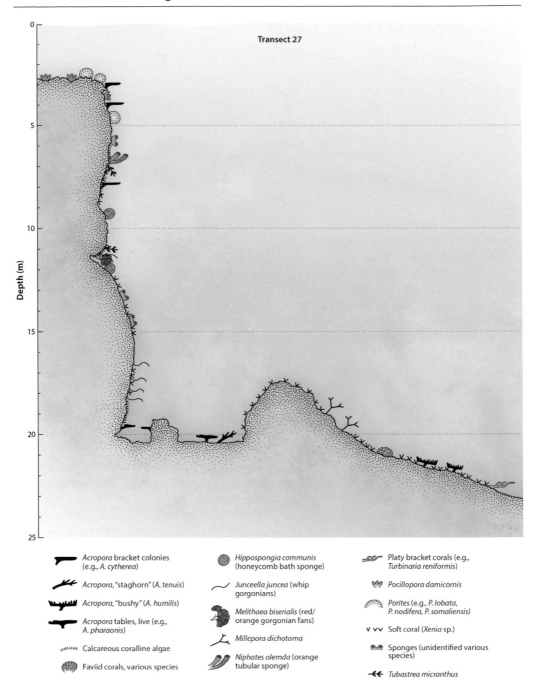

Transect 27

Depth (m)

Symbol	Description
	Acropora bracket colonies (e.g., *A. cytherea*)
	Acropora, "staghorn" (*A. tenuis*)
	Acropora, "bushy" (*A. humilis*)
	Acropora tables, live (e.g., *A. pharaonis*)
	Calcareous coralline algae
	Faviid corals, various species
	Hippospongia communis (honeycomb bath sponge)
	Junceella juncea (whip gorgonians)
	Melithaea biserialis (red/orange gorgonian fans)
	Millepora dichotoma
	Niphates olemda (orange tubular sponge)
	Platy bracket corals (e.g., *Turbinaria reniformis*)
	Pocillopora damicornis
	Porites (e.g., *P. lobata*, *P. nodifera*, *P. somaliensis*)
v vv	Soft coral (*Xenia* sp.)
	Sponges (unidentified various species)
	Tubastrea micranthus

Figure 25.1 Transect 27.

© Vine

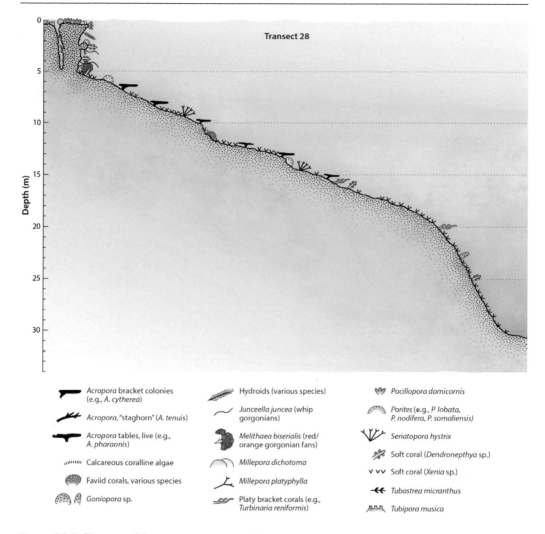

Figure 25.2 Transect 28.
© Vine

steeper sloping reef face from about 18–30 m. The latter is followed by a sloping soft sediment seabed.

Corals: *Tubipora musica*, *Millepora platyphylla*, and *Porites* were dominant on the reef top. At the reef crest, *P. damicornis* was very common. The undercut shallow reef face had a restricted coral fauna. However, from shaded crevices and underhangs, numerous *Junceella* and *Dendronephthya* occurred. On the main terrace (5–18 m), there were large *Acropora* tables and various other corals interspersed with patches of sand or *Xenia*-covered coralline rubble. *Seriatopora*, *Porites*, staghorn *Acropora*, and solitary forms such as *Fungia* and *Herpolitha* were abundant.

Figure 25.3 Scuba diver near propeller of *Umbria* wreck, Wingate reef, Red Sea, Sudan.
© WaterFrame/Alamy

Fish: Algal grazers such as *A. sohal*, *C. striatus*, and several parrotfish were abundant in the shallows and over the reef top. There was a slight long-reef current moving from north to south on the main reef slope, and planktivorous species faced this current. At approximately 13 m, there were shoals of *Chaetodon melannotus*. At 15 m, above the edge of the deep reef face, there were schools of *O. niger*. The general impression was that fish were relatively scarce and primarily concentrated in the upper 10 m.

Chapter 26

Sanganeb Reef

Sanganeb atoll lies 15 nautical miles east-northeast of Port Sudan. It is separated from Wingate reef and its northern extending ribbon reefs by about 7 nautical miles of open sea, averaging 800 m deep. The 50 m high lighthouse is a prominent landmark for navigation. Its large accommodation block and extensive ground area provide an ideal field station base, enabling authorised biologists to live at close quarters with their field studies. The sea around Sanganeb is always clear, and the biotopes are varied and flourishing. Species range from typically oceanic ones, such as large tuna, sailfish, hammerhead sharks, and the pelagic whitetip, *Carcharhinus longimanus*, to a wide range of reef-dwelling species and lagoonal forms such as the delicately coloured *P. flavivertex* together with many invertebrates more typical of lagoons or harbours than open-water coral reefs. There was a semi-resident school of dolphins, frequently sighted turtles, manta rays, schools of barracuda, and a variety of resident or migratory sharks and other large pelagic fish. Hussey et al. (2013) studied grey reef sharks at the nearby reef of Shaab Rumi. Peak shark observations were made when SST was 26–26.9°C and currents were strong. These conditions tended to occur during the winter/spring months (November to April), and the vast majority of sharks observed were female. Data were collected by recreational divers, and Hussey et al. recommended an expansion of citizen science for longer-term monitoring programmes.

In the winter/early spring months, especially November to April, scalloped hammerhead sharks form schools at the southwest and northeast points of Sanganeb atoll. These sharks may be observed at shallower depths, around 20 m, in the early morning (before 0900) or in the evening, immediately before sunset. At other times, they usually remain in deeper water, around 70–90 m or more. During May and June, sailfish are relatively common around the atoll, and they enter the shallow water over the top of the reef and in the lagoon.

There is a peak of zooplankton during January and February, when many of the planktivorous species appear to be most active. The north side of the atoll is exposed to predominantly northerly winds, and diving along this section is possible only during calm weather or when the wind is from the south. It is a fascinating reef, with flourishing coral growths, large shoals of fish, and dramatic steep profiles. Off the southwesterly point, the reef drops to around 30 m, and then an off lying pinnacle rises to about 20 m.

Above the pinnacle, in mid-water, there were large shoals of planktivorous fish. Grey reef sharks frequently cruise through the channel separating the pinnacle from the main reef, and here, in the early morning, one may observe large sharks. We have heard it described as one of the most magnificent diving locations in the world. Indeed, we would apply the same description to the whole of the Sanganeb reef.

Mergner and Schuhmacher (1985) published a quantitative analysis of Sanganeb's habitats.

DOI: 10.1201/9781003335795-30

Transect 29

Location: Situated on the south side of Sanganeb at approximately the southwest point of the atoll.

Figure 26.1 The long landing jetty at Sanganeb atoll.

© Vine

Figure 26.2 Sanganeb reef top.
© Vine

Morphology: It consists of a steep, reasonably smooth reef face, dropping from the surface to 60 m depth at an almost vertical angle. From the reef crest to 8 m depth, the reef is undercut. Numerous "chimneys" and caverns whose walls are covered by pink CCA interconnect, forming a network of channels at the reef edge.

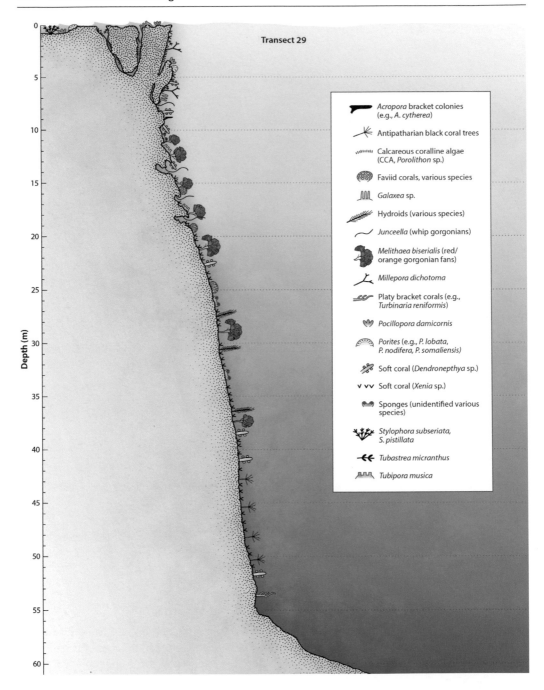

Figure 26.3 Transect 29.

© Vine

Corals: The shaded cliff face beneath the reef crest has solitary ahermatypic corals, including *Tubastrea* and *Balanophyllia*. Hermatypic corals are relatively scarce on the reef face. *Porites* and *Pocillopora* are present in shallow zones, while *Junceella* sp. (whip gorgonian), gorgonian fans, *D. savignyi*, *Xenia*, and antipatharians are characteristic of the deeper reef.

Fish: Fish were not abundant on the reef face, and the most frequently recorded species were *P. fridmani*, *P. squamipinnis*, and *P. dimidiatus*. At shallow depths, there were more crevices and live corals, which attracted more fish species.

Transect 30

Location: Situated about 20 m north of the southwest tip of Sanganeb reef.

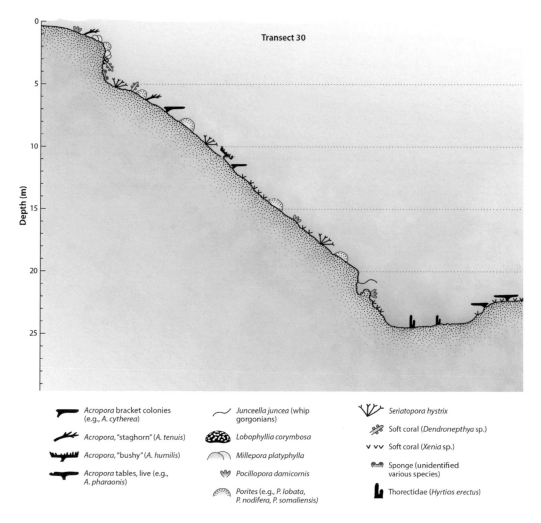

Figure 26.4 Transect 30 (Continued).

© Vine

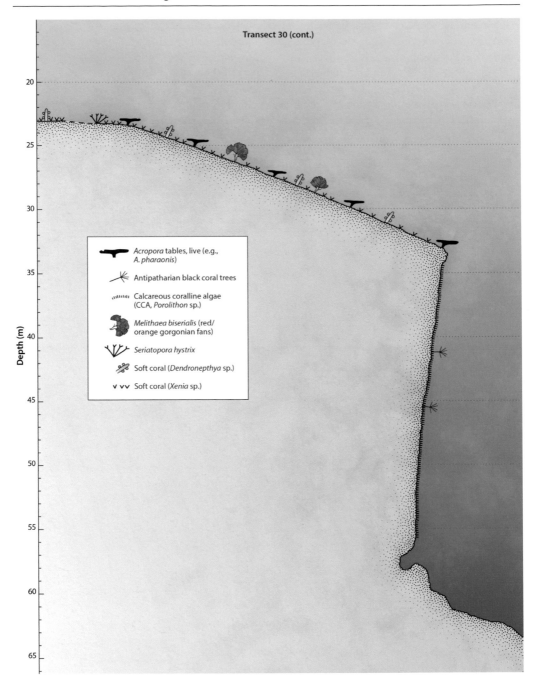

Figure 26.4 (Continued)

Morphology: It consists of a steeply inclined reef face from the crest to about 10 m, followed by a more gradual slope to 23 m. A more gently sloping reef terrace then follows a narrow sandy zone from 22 to 33 m. The 22-m terrace is foreshortened but extends horizontally for at least 50 m. From 34 m to about 60 m is a steep undercut reef wall.

The terrace at 22 m is a characteristic feature of the westerly-facing reef in this section of Sanganeb. As one moves towards the southwest point and from there along the southern edge of the reef, the terrace gradually diminishes in extent until it disappears altogether, and the reef face descends almost vertically from the crest to about 55 m.

Corals: The reef crest consisted of *M. platyphylla*, *Pocillopora*, and bushy *Acropora*. The shallow reef face had *Porites*, *M. dichotoma*, *Lobophyllia*, *Distichopora*, and many recently settled young colonies of *Dendronephthya* and sponges. The slope, which extends from 10 to 29 m, had an irregular surface, with colonies of *Stylophora*, *Porites*, *Acropora*, *Seriatopora*, *Pocillopora*, and a matting of *Xenia* on rock surfaces, particularly in the lower sections. *Acropora* tables were present on the terrace below the slope. This terrace at about 22 m had abundant *Xenia* and other soft corals. There were also numerous colonies of *Stylophora*. The gently sloping reef face from 22 to 33 m had occasional *Acropora* tables together with abundant soft corals. The steep undercut wall from 34 to 60 m had few corals growing on it. It was encrusted with CCA and had large black coral trees attached to it. Unfortunately, these were being destroyed by divers searching for trophies.

Fish: On the main terrace, around 25 m or off the slope to seaward, one could observe both scalloped and great hammerhead sharks together with other shark species in the early morning or late afternoon, especially during the winter months. At this location, a Spanish diver filmed a shoal of tuna attacking a single shark by nose bumping in a similar way to that reported for dolphins attacking sharks. Field records from the present study were partially reported by Vine and Vine (1980), who compiled a list of species.

Chapter 27

Shaab Rumi

Frequency and distribution of corals on the underwater garage at Shaab Rumi are plotted in a series of diagrams. Coral growth on Shaab Rumi was prolific and diverse, and it, in turn, supported a rich assemblage of reef fishes. Bright red and orange *Pseudanthias*, and their mimics, *Ecsenius midas*; planktivorous damsel fish such as *Chromis caerulea* and larger grazers like *Zebrasoma veliferum*, the parrotfish *Hipposcarus harid, Chlorurus sordidus,* wrasse like *Thalassoma lunare*, and the goatfish, *Parupeneus macronemus*. created a kaleidoscope of colours on the reef and around the garage. But reefs were already in decline, with Cousteau's team alarmed by the absence of larger fish such as groupers and snappers – prime targets for fishermen.

My own fieldwork at Shaab Rumi in 1974–1976 continued to record coral growth rates, fish distribution, and key stress factors at this atoll reef. Its designation as "Roman Reef" was reinforced by the find at this location of an ancient water or wine jar, typical of Roman times.

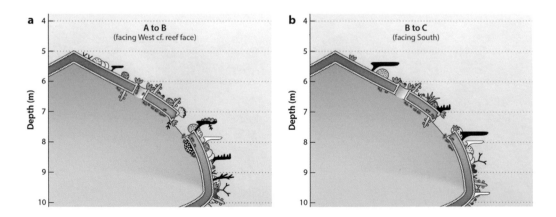

Figure 27.1 Transects of Cousteau's underwater garage in 1974–1975. (*a*) Side A to B, facing approximately due west. (*b*) Side B to C, facing approximately south. (*c*) Side C to D, facing approximately east. A table of *Acropora pharaonis* extended 39 cm in length in 10 months. (*d*) Side D to A, facing north. Illustrations by Fiona Martin, based on notes and photographs labelled by the author (Continued).

© Vine

DOI: 10.1201/9781003335795-31

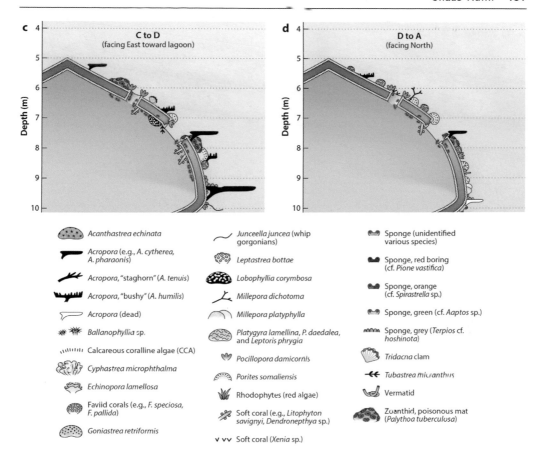

c   C to D
(facing East toward lagoon)

d   D to A
(facing North)

Depth (m)

Acanthastrea echinata

Acropora (e.g., A. cytherea,
A. pharaonis)

Acropora, "staghorn" (A. tenuis)

Acropora, "bushy" (A. humilis)

Acropora (dead)

Ballanophyllia sp.

Calcareous coralline algae (CCA)

Cyphastrea microphthalma

Echinopora lamellosa

Faviid corals (e.g., F. speciosa,
F. pallida)

Goniastrea retriformis

Junceella juncea (whip
gorgonians)

Leptastrea bottae

Lobophyllia corymbosa

Millepora dichotoma

Millepora platyphylla

Platygyra lamellina, P. daedalea,
and Leptoris phrygia

Pocillopora damicornis

Porites somaliensis

Rhodophytes (red algae)

Soft coral (e.g., Litophyton
savignyi, Dendronepthya sp.)

v vv   Soft coral (Xenia sp.)

Sponge (unidentified
various species)

Sponge, red boring
(cf. Pione vastifica)

Sponge, orange
(cf. Spirastrella sp.)

Sponge, green (cf. Aaptos sp.)

Sponge, grey (Terpios cf.
hoshinota)

Tridacna clam

Tubastrea micranthus

Vermatid

Zoanthid, poisonous mat
(Palythoa tuberculosa)

Figure 27.1 (Continued)

Chapter 28

Discussion

This study was established to record the main characteristics of some diverse areas of coral reefs, approximately 50 years ago. I recognised the need for some baseline studies that would record descriptions and data on corals and fish but had not envisaged how valuable the data would turn out to be. Climate change, global warming, acidification, increased sedimentation, overfishing, and imbalances leading to phase shifts from coral domination to algal domination, were all part of a much bigger story than any of us had predicted. Alarm bells had rung on the Great Barrier Reef, where massive aggregations of CoTS were transforming vibrant reefs into graveyards and rubble fields, and nobody was sure why this was happening.

Anthropogenic disturbance was generally close to population centres and reef sites regularly visited by tourists. Among the prime dive sites in the Red Sea were the remains of Cousteau's Conshelf 2 experiment on Shaab Rumi, where we plotted the settlement and growth of a number of coral species.

Four years after the Conshelf 2 project, in 1968, Cousteau's team returned to the historic site and noted the opportunity it presented to accumulate data on settlement and growth of key species. A profusion of corals, algae, and sponges garlanded the garage. Seven years later, we established our own underwater survey at the Shaab Rumi site, where we plotted some of the fastest growth rates recorded by reef-building corals.

Colonisation of marine life was frequently initiated by the settlement of *Porites*, *Pocillopora*, and *Acropora* spp., each of which relies on herbivores such as fish, to control algae and sponges, maintaining a healthy mix of habitats. However, nothing is permanent. These coral habitats are in a state of flux, often resulting in a shift towards brown algae or attacks by sponges and other invertebrates. As this happens, primary colonisers such as sponges and coral may die off, exposing new surfaces for the settlement of some slower-growing species.

Physical reef morphology is critical in influencing the wide range of habitats on living reefs. For example, the steep drop-offs of the fringing reef near Port Sudan, or at the southern rim of Sanganeb atoll, provide shade, near-reef currents, access to plankton, and relative protection from sediments that might otherwise clog the tentacles of filter feeders. Meanwhile, flooded inlets (wadis) along the coast, such as at Suakin and Arikiyai, are endowed with rich communities of sand and mud dwellers, along with sponges and soft corals.

Reef terraces, where steeply inclined reef-faces level out for a few metres before descending once again towards the dark blue depths, provide intermediate habitats that were originally formed as ancient shores during previous Ice Ages (c. 12,000 years ago), when sea levels were much lower than at present (Neumann & Moore 1975). We need look no further for evidence of change. The Red Sea has been described as an "ocean in the making" and has been in a state of flux for millennia.

DOI: 10.1201/9781003335795-32

Figure 28.1 A scuba diver investigates Jacques Cousteau's Conshelf 2 at Shaab Rumi.
© Reinhard Dirscherl/Alamy

Figure 28.2 Coral reefs destroyed by outbreaks of the crown-of-thorns starfish, *Acanthaster planci*.
© Mike Veitch/Alamy

Some of the interactions triggered by humankind and nature that are described earlier have drawn the attention of scientists since minor changes can create massive impacts on reefs. Thus, for example, too much fishing can turn a diversified coral reef into a monotonous, algae-dominated one with restricted biodiversity. A "plague" of starfish can trigger a shift from corals to algae as well as sound the death knell for many coral-dependent fish. A rise in effluent pollution or a rusting shipwreck can boost the growth of killer sponges that cover large areas of corals, leaving them dead and rotting. A small increment in sea temperatures can trigger coral bleaching and widespread coral mortalities. The demise of coral reefs results in a catastrophic reduction in carbon sequestration and a consequential acidification of coastal waters, with implications for the survival of organisms whose larvae, juveniles, and adults are dependent on calcium carbonate.

As I noted at the beginning of this book, life in coral seas was changing, and changing fast. Along with the loss of reef-building corals, we were witnessing a widespread phase shift – from corals to macroalgae – accompanied by the disappearance of many reef fish and their related flora and fauna.

If we are to have any chance of managing change on reefs such as those of the Red Sea, we must first understand these processes at work. Baseline studies open doors to multidisciplinary research on a host of species, some of which can be directly beneficial to our own survival

Figure 28.3 Corals, sponges, and algae coexist in a delicate balance to build reefs that provide homes for a myriad of fish and invertebrates. Reefs help protect shorelines from erosion, support thriving tourist industries, and provide valuable resources to local and scientific communities.

© Hans Sjöholm

(Klaus et al. 2008; Beyer et al. 2015a, 2015b; Nasr 2015). We need corals and sponges, not just for their essential roles in building and maintaining sustainable reefs but also for their roles as living laboratories for the biopharma industries that are finding unique chemicals with proven medicinal properties (Abdelhameed et al. 2020).

We don't just need corals to survive, however. We need them to thrive and adapt to a changing world. To do that, we need to look into their own complex and competitive biospheres. The relations between key species on coral reefs: fish, hydrozoans, soft and hard corals, sponges, other invertebrates, turf algae, macroalgae, CCA, bacteria, and seagrasses, and especially the vital role that fish play in reef ecology, are becoming clearer to marine scientists.

Reduction of fish populations usually triggers a reduction of live corals (Loh et al. 2015), while loss of corals generally has the same negative impact on coral reef fish species. But which comes first? Either eventuality is likely to involve a reduction in biodiversity and a shift to algal dominance. Many studies have highlighted the importance of balanced ecosystems in which both fish and corals play complementary roles (Wilson et al. 2006; Haas et al. 2010).

Wilson and colleagues (2006) undertook a desk study of documents recording disturbances of coral and associated losses of reef fishes. A more than 10 percent decline in coral cover was accompanied by a 60 percent drop in local fish diversity within three years of the disturbances. The most impacted species were those reliant on live coral for food and shelter. Cataclysmic

Figure 28.4 A largely dead *Acropora* table being overgrown by algae. Associated fish are greatly diminished despite the existence of complex scaffolding and numerous hiding places.

© Martin Habluetzel/Alamy

disruptions resulting in widespread coral mortality (such as cyclone damage or *Acanthaster* outbreaks) have a more significant impact on fishes from all trophic levels than disturbances that kill corals but leave the reef framework intact (Wilson et al. 2006).

Studies on *A. planci* and its ecological impact on coral reefs (Vine 1970, 1973, 2019; Ormond et al. 1990; Pratchett 2001; Yuasa et al. 2017; Wilmes 2018) have provided insights into mass coral mortalities caused by other species, including the killer sponge *T. hoshinota*. In all mass killings, there were associated dramatic reductions in the diversity of reef fishes.

Studies of fish distribution, behaviour, and population dynamics illustrate, time and time again, that the greater the complexity of tropical marine ecosystems, the greater their resilience to stresses such as overfishing, disease, pollution, and rise in SST.

There are few ecosystems where the battle for space to settle and grow is more intense and ubiquitous than coral reefs. This is particularly apparent on Red Sea reefs where highly special-ised habitats were dominated by a few prominent species or where reef residents may range over the more extensive coral domain.

Melbourne-Thomas et al. (2011) interrogated the complex structure of reef dynamics that modulates the battle for space on the reef, creating a model that simulates the ecological pro-cesses at play over large reef areas. The shift from coral-dominated to macroalgae-dominated habitats on reefs can be triggered by factors other than just overfishing of herbivorous fish (Arias-González et al. 2017). Sedimentation due to coastal development or eutrophication due to pollution and runoff also play critical roles in destabilising coral reefs and promoting their replacement by algal habitats.

Dungonab Bay and its environs provided interesting examples of these phenomena. The exposed outer and sheltered inner reaches of the bay differ noticeably from each other, display-ing individual responses to stress factors such as rises in SST, coral bleaching, acidification, and aggregations of *Acanthaster*.

Cataclysmic mortality of corals was more widespread among those species associated with open reefs than in cut-off embayments such as inner Dungonab Bay, where the diversity of spe-cies was severely restricted to almost 100 percent *Galaxea* colonies that had clearly survived such episodes in the past. They appear to have built up some resistance to coral bleaching while the CoTS diet switched to the soft coral *Xenia*, something rarely observed among healthy reefs in more exposed locations (Yuasa et al. 2017).

In the early 1970s, when much of the fieldwork for this study took place, Sudan's coral reefs were widely reported as being healthy, with diverse habitats that attracted biologists, sport divers, and fishers. Amid growing concern that such a pristine environment should be protected for future generations, efforts were discussed to preserve the status quo and prevent the destruc-tion of the coral reef communities.

Documentary films like *Chasing Coral* (directed by Jeff Orlowski released in 2017 by Netflix and YouTube) drew emotional responses from those whose lives had been spent studying the Blue Planet's most diverse marine habitats. Coral reefs indeed face a bleak future, but one that is not entirely without hope. Interventions to assist coral reefs that are changing on a global scale were discussed by Anthony et al. (2020). They proposed architectural cooling and shading that would promote coral recruitment. Additionally, Grottoli et al. (2014) studied the cumulative impact of the annual event of coral bleaching, which can turn some winners into losers, empha-sising the long-term legacy of climate change on coral–algal reefs, especially those living close to peaks in SST.

Meanwhile, Gajdzik et al. (2021) discuss a portfolio of climate-tailored approaches to advance the design of MPAs in the Red Sea. They suggest that, while MPAs, such as the

Sanganeb-Dungonab World Heritage Site, were located with the best intentions, it is unclear to what extent wildlife is benefitting or how effective their management has been. Their central issue with the process was that the MPAs were not designated on the basis of climate considerations. They propose working with nature in the Red Sea rather than against it, taking a new look at the issue by tailoring conservation measures to enhance climate change mitigation and adaptation. A more climate-centred approach would:

1 Take account of coral bleaching susceptibility, producing "a more resilient network of MPAs by safeguarding reefs from different thermal regions that vary in spatiotemporal bleaching responses, thus lowering the risk that all protected reefs will bleach simultaneously";
2 Preserve the "basin-wide genetic connectivity patterns" that are assisted by mesoscale eddies, maintaining species potential to adapt to environmental changes;
3 Protect mangrove forests in the northern and southern Red Sea that act as major carbon sinks, helping to offset greenhouse gas emissions.

A vision of what scientists expect will occur in the Red Sea (see Kleinhaus et al. 2020) was provided by events on the far northern Great Barrier Reef in Australia, where, in 2016, heat stress bleached 85 percent of reefs, killing 29 percent of the reef's shallow-water corals. Bleaching also occurred in parts of the western Indian Ocean and central Red Sea), including a 50 percent death toll in Seychelles. From 2014 to 2017, more than 75 percent of global reefs suffered mass bleaching, killing 30 percent of reef-building corals at shallow depths – the most destructive event of its kind on record (Loch et al. 2004; Lewis & Mallela 2018). Despite such a bleak prognosis – with 70–90 percent of reef-building corals expected to die by mid-century – there are some glimmers of hope, provided by corals such as *S. pistillata* in the northern Red Sea, where certain corals have adapted to withstand temperature increases of as much as 7°C (Evensen et al. 2021).

Utilising the simulator at the Australian Institute of Marine Science (AIMS), researchers demonstrated that a rise in temperatures is accompanied by a slowdown in growth rates. Fast-growing species such as *Acropora* tables are also among the most sensitive to stress, bleaching more readily than others and being less resilient against storms. Given that reef regeneration is crucially dependent on such species (i.e., table Acroporas), this suggests that they are optimally adapted to support regrowth. According to Dr. Juan Ortiz, senior author of the study, they may have "limited potential for adaptation to future hotter conditions." After exposing selected species to a set of temperatures ranging from 19 to 31°C, the team was surprised by the uniformity of response. "While individuals of the same species grew at vastly different rates, the temperature at which they grew fastest was remarkably similar."

"The world's third-largest coral reef system historically suffered from severe mass bleaching only when exposed to both unusually high temperature and excess nutrients" (DeCarlo 2020). Massive pollution events, such as those in the Gulf during the Iraq-Kuwait War, remind us of the extraordinary resilience of tropical marine species (Krupp & Müller 1994). It is remarkable that critically damaged habitats have recovered as well as they have, but the situation has been exacerbated by climate change. The jury may be out on the question, "Can we do anything about it?" (Sheppard 2016), but we should take some comfort from discoveries of physiological adaptations whereby some species demonstrate unusual abilities to survive high-temperature exposures (McCook 2009).

In some ways, the relative slowdown in coastal economic development has been good news for the marine environment, but this is unlikely to continue. Coral reefs of 50 years ago may

Figure 28.5 Pocillopora sp. shows remarkable resilience while other corals are bleaching nearby.
© Hans Sjöholm

seem to be a long-lost dream, but neither their beauty nor their economic potential has been lost altogether. Offshore reefs such as Sanganeb, Shaab Rumi, Wingate, and Towartit still offer some of the most attractive dive sites in the world. Is it too much to hope that the next half century will turn the tide back in favour of coral survival?

References

Abdelhameed RF et al. (2020) New cytoxic natural products from the Red Sea sponge *Stylissa carteri*. *Marine Drugs* 18(5): 241

Abecasis D et al. (2013) Herbivores strongly influence algal recruitment in coral- and algal-dominated coral reef habitats. *Marine Ecology Progress* series 486: 153–164

Abelson A (2020) Are we sacrificing the future of coral reefs on the altar of the "climate change" narrative? *ICES Journal of Marine Science* 77(1): 40–45

Aeby GS et al. (2017) First record of crustose coralline algae diseases in the Red Sea. *Bulletin of Marine Science* 93(4)

Aerts L (2000) Dynamics between standoff interactions in three reef sponge species and the coral *Montastraea cavernosa*. *Marine Ecology* 191–204: 213–214

Aerts L et al. (1997) Quantification of sponge/coral interactions In a physically stressed reef community, NE Columbia. *Marine Ecology Progress Series* 148(1–3): 125–134

Aerts L & Van Soest R (1997) Quantification of sponge/coral interactions in a physically stressed reef community, NE Columbia. *Oceanographic Literature Review* 9(44): 987

Ahmed AG (2015) Correlation between physical and chemical parameters and marine macro zooplankton community around Port Sudan area. *Journal of Marine Biology & Oceanography* 4(2)

Aini SN (2021) Monthly progression rates of the coral-killing sponge *Terpios hoshinota* in Sesoko Island, Okinawa, Japan. *Coral Reefs* 40(3): 973–981

Anthony KR et al. (2020) Interventions to help coral reefs under global change – A complex decision challenge. *PLoS ONE* 15(8)

Antonius A (1988a) Distribution and dynamics of coral diseases in the Eastern Red Sea. *Proceedings of the 6th International Coral Reef Symposium*, Townsville 2: 293–298

Antonius A (1988b) Black band disease behavior on Red Sea reef corals. *Proceedings of the 6th International Coral Reef Symposium*, Townsville 3: 145–150

Arias-González JE et al. (2017) A coral-algal phase shift in Mesoamerica not driven by changes in herbivorous fish abundance. PLoS ONE 12(4): e0174855

Aronson, RB et al. (2003). Causes of coral reef degradation. *Science* 302(5650): 1502–1504

Aswani S et al. (2015) Scientific frontiers in the management of coral reefs. *Frontiers in Marine Science* 2: 50

Ateweberhan, MJH et al. (2016) Effects of extreme seasonality on community structure and functional group dynamics of coral reef algae in the southern Red Sea (Eritrea). *Coral Reefs* 25: 391–406

Ayling A (2023) Video of 40 years observations of corals on Snapper Reef. *YouTube*. https://youtu.be/e6JCQmN6_S8

Barott KL et al. (2012) Microbial to reef scale interactions between the reef-building coral *Montastraea annularis* and benthic algae. *Proceedings of the Royal Society B* 279: 1655–1664

Barshis DJ et al. (2013) Transcriptomics of coral climate change resilience. *Molecular Ecology* 24(7): 1467–1484

Bell JJ et al. (2013) Could some coral reefs become sponge reefs as our climate changes? *Global Change Biology* 19(9): 2613–2624

Benayahu Y & Loya Y (1981) Competition for space among coral-reef sessile organisms at Eilat Red Sea. *Bulletin of Marine Science* 313: 514–522.

Berumen ML et al. (2013) The status of coral reef ecology research in the Red Sea. *Coral Reefs* 32(3): 737–748

Berumen ML et al. (2019) *Corals of the Red Sea Coral Reefs of the Red Sea.* Gland: IUCN, pp. 123–155

Beyer J et al. (2015a) *Marine Ecological Baselines and Environmental Impact Assessment Studies in the Sudanese Coastal Zone.* Jeddah: PERSGA (Regional Organization for the Conservation of the Environment of the Red Sea and Gulf of Aden)

Beyer J et al. (2015b) *Guideline for Environmental Monitoring in Sudanese Marine Waters in Connection with Offshore Oil and Gas Industry Activities.* Jeddah: PERSGA (Regional Organization for the Conservation of the Environment of the Red Sea and Gulf of Aden)

Birrell CL et al. (2008) Effects of benthic algae on the replenishment of corals and the implications for the resilience of coral reefs. In *Oceanography and Marine Biology* (pp. 31–70). CRC Press

Bogorodsky SV & Randall JE (2019) Endemic fishes of the Red Sea. In *Oceanographic and Biological Aspects of the Red Sea* (pp. 239–265). Springer

Bryan PG (1973) Growth rate toxicity and distribution of the encrusting sponge *Terpios sp* (Hadromerida: Suberitidae) in Guam Mariana Islands. *Micronesica* 9: 237–242

Buddemeier RW & Kinzie III RA (1976) Coral growth. In Harold Barnes (Ed.), *Oceanography and Marine Biology: An Annual Review* (vol. 14, pp. 183–225). Aberdeen: Aberdeen University Press

Cacciapaglia C & van Woesik R (2015) Reef-coral refugia in a rapidly changing ocean. *Global Change Biology* 216: 2272–2282.

Camp EF et al. (2018, February 2) The future of coral reefs subject to rapid climate change: Lessons from natural extreme environments. *Frontiers in Marine Science*

Carpenter RC et al. (2006) Local and regional scale recovery of *Diadema* promotes recruitment of scleractinian corals. *Ecology Letters* 93: 271–280.

Ceccarelli DM et al. (2006) Impacts of simulated overfishing on the territoriality of coral reef damselfish. *Marine Ecology Progress Series* 309: 255–262

Chisholm JRM (2003) Primary productivity of reef-building crustose coralline algae. *Limnol Oceanogr* 48: 1376–1387

Clark E, Pohle JF & Halstead B (1998) Ecology and behavior of tilefishes, *Hoplolatilus starcki, H. fronticinctus* and related species (Malacanthidae): Non-mound and mound builders. *Environmental Biology of Fishes* 52(4): 395–417

Cousteau J-Y (1953) *Silent World* (148pp). London: Hamish Hamilton

Cousteau J-Y (1971) *Life and Death in a Coral Sea* (302pp). London: Cassell

Crossland C (1919) Red Sea province dangers of pearl diving. *Sudan Notes and Records* 2(3): 234–236

Crossland C (1956) The cultivation of the mother-of-pearl oyster in the Red Sea. *Marine and Freshwater Research* 82: 111–130.

Crylen J (2018) Living in a world without sun: Jacques cousteau homo aquaticus and the dream of dwelling undersea. *JCMS: Journal of Cinema and Media Studies* 581: 1–23.

Cziesielski MJ et al. (2021) Investing in blue natural capital to secure a future for the Red Sea ecosystems. *Frontiers in Marine Science* 7: 1183

Dang VDH et al. (2020) Grazing effects of sea urchin *Diadema savignyi* on algal abundance and coral recruitment processes. *Scientific Reports* 10(1): 1–9

Das RR (2020) Incursion of the killer sponge *Terpios hoshinota* Rützler & Muzik 1993 on the coral reefs of the Lakshadweep archipelago. *Arabian Sea Journal of Threatened Taxa* 12(14): 17009–17013

da Silva RF et al. (2020) Modelling three-dimensional flow over spur-and-groove morphology. *Coral Reefs* 39(6): 1841–1858

DeCarlo TM (2020) Nutrient-supplying ocean currents modulate coral bleaching susceptibility. *Science Advances* 6(34): eabc5493

DeVantier L et al. (2000a) Coral communities of the central-northern Saudi Arabian Red Sea. *Fauna of Arabia* 18: 23–66. Published online 16 November 2011

DeVantier L et al. (2000b, February 5th–9th) Coral bleaching in the central northern Saudi Arabian Red Sea, August-September 1998. In A Abduzinada, E Joubert, & F Krupp (Eds.), *Proceedings of an International Symposium on the Extent and Impact of Coral Bleaching in the Arabian Region* (pp. 110–127). National Commission for Wildlife Conservation and Development, Riyadh, Saudi Arabia

De Voogd NJ et al. (2013) The coral-killing sponge *Terpios hoshinota* invades Indonesia. *Coral Reefs* 323: 755–755

Diaz-Pulido G et al. (2010) The impact of benthic algae on the settlement of a reef-building coral. *Coral Reefs* 29(1): 203–208

Eakin CM et al. (2019) The 2014–2017 global-scale coral bleaching event: Insights and impacts. *Coral Reefs* 38(4): 539–545

Edmunds PJ & Carpenter RC (2001) Recovery of *Diadema antillarum* reduces macroalgal cover and increases abundance of Juvenile Corals on a Caribbean Reef. *Proceedings of the National Academy of Sciences – PNAS* 989: 5067–5071.

Edwards AJ & Head SM, (eds) (1987) *Red Sea (Key Environments)*. Oxford: Published in Collaboration with the International Union for Conservation of Nature and Natural Resources by Pergamon Press, 441pp. ISBN 008 028873 1

Edwards AJ & Rosewell J (1981) Vertical zonation of coral reef fishes in the sudanese red sea. *Hydrobiologia* 791: 21–31.

El Hag AD (1999). Biodiversity assessment, threats and sustainability of marine ecosystems in Sudanese Red Sea coast. *A Report for the National Biodiversity Strategy and Action Plan*. Red Sea University Khartoum Sudan

Elliott JP (2016) Morphological plasticity allows coral to actively overgrow the aggressive sponge *Terpios hoshinota* (Mauritius Southwestern Indian Ocean). *Marine Biodiversity* 46(2): 489–493

Elliott JP et al. (2016) How does the proliferation of the coral-killing sponge *Terpios hoshinota* affect benthic community structure on coral reefs? *Coral Reefs* 35(3): 1083–1095

Elsheikh BA et al. (2018) Seasonal variations of hydrographic parameters off the Sudanese coast of the Red Sea, 2009–2015. *Regional Studies in Marine Science* 18: 1–10. Elsevier

Eltayeb Ali et al. (2016) Coral diversity and similarity along sudanese Red Sea fringing reef. *International Journal of Advanced Research* (Indore) 44: 720–727.

Evensen NR et al. (2021) Remarkably high and consistent tolerance of a Red Sea coral to acute and chronic thermal stress exposures. *Limnology and Oceanography* 66(5): 1718–1729

Fidelman P et al. (2019) Regulatory implications of coral reef restoration and adaptation under a changing climate. *Environmental Science & Policy* 100: 221–229

Fine M et al. (2019) Coral reefs of the Red Sea – Challenges and potential solutions. *Regional Studies in Marine Science* 25: 100498

Florian R et al. (2021) High summer temperatures amplify functional differences between coral- and algae-dominated reef communities. *The Bulletin of the Ecological Society of America* 102(1)

Fong P & Paul VJ (2011) Coral reef algae. In Z. Dubinsky & N. Stambler (Eds.), *Coral Reefs: An Ecosystem in Transition*. Dordrecht: Springer. https://doi.org/10.1007/978-94-007-0114-4_17

Foster N (2005) Patch dynamics of coral reef macroalgae under chronic and acute disturbance. *Coral Reefs* 24: 681–692. http://doi.org/10.1007/s00338-005-0058-5

Fujii T et al. (2011) Coral-killing cyanobacteriosponge (*Terpios hoshinota*) on the Great Barrier Reef. *Coral Reefs* 30(2): 483–483

Furby KA (2013) Susceptibility of central Red Sea corals during a major bleaching event. Coral Reefs 32(2): 505–513

Gajdzik L et al. (2021) A portfolio of climate-tailored approaches to advance the design of marine protected areas in the Red Sea. *Global Change Biology* 21(17): 3956–3968

González-Rivero M et al. (2011) The role of sponge competition on coral reef alternative steady states. *Ecological Modelling* 222(11): 1847–1853

Greenlaw JP (1976) The coral buildings of Suakin. In *Islamic Architecture Planning Design and Domestic Arrangements in Red Sea Port* (1995 ed.). London and New York: Paul Kegan International

Grottoli AG et al. (2014) The cumulative impact of annual coral bleaching can turn some coral species winners into losers. *Global Change Biology* 20(12): 3823–3833

Haas A et al. (2010) Seasonal monitoring of coral – Algae interactions in fringing reefs of the Gulf of Aqaba Northern Red Sea. *Coral Reefs* 29(1): 93–103

Harborne AR et al. (2017) Multiple stressors and the functioning of coral reefs. *Annual Review of Marine Science* 9: 445–468

Harrington L et al. (2004) Recognition and selection of settlement substrata determine post-settlement survival in corals. *Ecology* 85(12): 3428–3437

Hass H (1952) *Under the Red Sea*. London: Jarrolds, 208 pp

Hass H (1971) *Challenging the Deep*. New York: William Morrow, 265 pp

Hass H (1975) *Conquest of the Underwater World:* Devon: David & Charles, 407pp

Hata H (2002) Effects of habitat-conditioning by the damselfish *Stegastes nigricans* (Lacepède) on the community structure of benthic algae. *Journal of Experimental Marine Biology and Ecology* 280: 95–106

Hata H & Kato M (2004) Monoculture and mixed-species algal farms on a coral reef are maintained through intensive and extensive management by damselfishes. *Journal of Experimental Marine Biology and Ecology* 313(2): 285–296.

Hata H et al. (2012) Geographic variation in the damselfish-red alga cultivation mutualism in the Indo-West Pacific. *BMC Evolutionary Biology* 10: 185.

Haywood MDE (2019) Crown-of-thorns starfish impede the recovery potential of coral reefs following bleaching. *Marine Biology* 166(7): 1–15

Hazraty-Kari S et al. (2021) Baseline assessment of coral diseases in an environmentally extreme environment of the northern Persian Gulf. *Marine Pollution Bulletin* 171: 112707

Head SM (1980) *The Ecology of Corals in the Sudanese Red Sea*. Ph.D. thesis. University of Cambridge

Head SM (1987) Corals and coral reefs of the Red Sea. Chapter 7. In AJ Edwards & SM Head (Eds.), *Red Sea (Key Environments)* (pp. 128–151). Oxford: Published in Collaboration with the International Union for Conservation of Nature and Natural Resources by Pergamon Press

Hodgsonn G & Liebeler J (2002) The global coral reef-5 years of reef check. *Institute of the Environment*. Los Angeles: Reef Check Foundation LA

Hoeksema BW et al. (2014) Partial mortality in corals overgrown by the sponge *Terpios hoshinota* at Tioman Island, Peninsular Malaysia (South China Sea). *Bulletin of Marine Science* 90(4): 989–999

Hussey NE et al. (2013). SCUBA diver observations and placard tags to monitor grey reef sharks, Carcharhinus amblyrhynchos, at Sha'ab Rumi, The Sudan: Assessment and future directions. *Journal of the Marine Biological Association of the United Kingdom* 93(2): 299–308

Jones RJ & Hoegh-Guldberg IO (2001) Diurnal changes in the photochemical efficiency of the symbiotic dinoflagellates (Dinophyceae) of corals: Photoprotection, photoinactivation and the relationship to coral bleaching. *Plant, Cell & Environment* 24: 89–99

Kamenos NA & Hennige SJ (2020) Commentary: Commentary on reconstructing four centuries of temperature-induced bleaching on the great barrier reef by hoegh-guldberg et al 2019 and DeCarlo 2020. *Frontiers in Marine Science* 7: 750

Kattan A (2017) Reef fish communities in the central red sea show evidence of asymmetrical fishing pressure. *Marine Biodiversity* 474: 1227–1238.

Keller C (1889) Die spongien Fauna des Rothen Meeres. Zeitschr. wiss. *Zool* 48: 311

Kessel ST et al. (2017) Conservation of reef manta rays (Manta alfredi) in a UNESCO World Heritage Site: *Large-scale island development or sustainable tourism? PLoS ONE* 12(10): e0185419

Klaus R (2015) Coral reefs and communities of the central and southern Red Sea (Sudan Eritrea Djibouti and Yemen). In N Rasul & I Stewart (Eds.), *The Red Sea Springer Earth System Sciences*. Berlin and Heidelberg: Springer. https://doiorg/101007/978-3-662-45201-1_25

Klaus R et al. (2008) Ecological patterns and status of the reefs of Sudan. *Proceedings of the 11th International Coral Reef Symposium*, Ft. Lauderdale, FL, 7–11 July 2008 Session Number 18

Kleinhaus K et al. (2020) Science diplomacy and the Red Sea's unique coral reef: It's time for action. *Frontiers in Marine Science* 7: 90

Kötter I (2001) Endoscopic exploration of red sea coral reefs reveals dense populations of cavity-dwelling sponges. *Nature* (London) 413(6857): 726–730.

Krupp F & Müller T (1994) The status of fish populations in the northern Arabian Gulf two years after the 1991 Gulf War oil spill. *Courier Forschungsinstitut Senckenberg* 166: 67–75

Kumagai NH et al. (2018) Ocean currents and herbivory drive macroalgae-to-coral community shift under climate warming. *Proceedings of the National Academy of Sciences* 115(36): 8990–8995

Lang JC & Chornesky EA (1990) Competition between scleractinian reef corals – A review of mechanisms and effects. In Z Dubinsky (Ed.), *Ecosystems of the World: Coral Reefs* (pp 209–252). Amsterdam: Elsevier

Langmead O & Chadwick-Furman N. (1999) Marginal tentacles of the corallimorpharian *Rhodactis rhodostoma*. 1. Role in competition for space. *Marine Biology* 134: 479–489. https://doi.org/10.1007/s002270050564

Lewis S & Mallela J (2018) A multifactor risk analysis of the record 2016 Great Barrier Reef bleaching. *Bulletin of the American Meteorological Society* 991: S144–S149.

Liao M et al. (2007) The "black disease" of reef-building corals at Green Island, Taiwan – outbreak of a cyanobacteriosponge, *Terpios hoshinota* (Suberitidae: Hadromerida). *Zoological Studies* 46: 520

Lieske E & Myers R (2004) *Coral Reef Guide Red Sea*. London: Collins

Lin Z, et al. (2017) Transcriptome profiling of *Galaxea fascicularis* and its endosymbiont *Symbiodinium* reveals chronic eutrophication tolerance pathways and metabolic mutualism between partners. *Scientific Reports* 7(1): 1–14

Loch K et al. (2004) Coral recruitment and regeneration on a Maldivian reef four years after the coral bleaching event of 1998 Part 2: 2001–2002. *Marine Ecology* 25(2): 145–154

Loh TL et al. (2015) Indirect effects of overfishing on Caribbean reefs: Sponges overgrow reef-building corals. *PeerJ* 3: e901c

Loya Y (1976) The recolonization of Red Sea corals affected by natural catastrophes and man-made perturbations. *Ecology* 57: 278±289

MacDonald C (2016) Depth, bay position and habitat structure as determinants of coral reef fish distributions: Are deep reefs a potential refuge? *Marine Ecology Progress Series* 561 217–231

Madduppa H (2017) Persistent outbreaks of the "black disease" sponge *Terpios hoshinota* in Indonesian coral reefs. *Marine Biodiversity* 47(1): 149–151

Manasrah R et al. (2019) Physical and chemical properties of seawater in the Gulf of Aqaba and Red Sea. In *Oceanographic and Biological Aspects of the Red Sea* (pp. 41–73). Cham: Springer

Manasrah R et al. (2020) Physical and chemical properties of seawater during 2013–2015 in the 400 m water column in the northern Gulf of Aqaba Red Sea. *Environmental Monitoring and Assessment* 192(3): 1–16

McCook LJ (2009) Management under uncertainty: Guide-lines for incorporating connectivity into the protection of coral reefs. *Coral Reefs* 28(2): 353–366

McKenna SA et al. (1997) Interactions between the boring sponge *Cliona lampa* and two hermatypic corals from Bermuda. In *Proceedings of the 8th International Coral Reef Symposium* (Vol. 2, pp. 1369–1374). Bermuda: Bermuda Biological Station for Research, Ferry Reach GEO1

Melbourne-Thomas JJ et al. (2011) Regional-scale scenario modelling for coral reefs: A decision support tool to inform management of a complex system. *Ecological Applications* 21(4): 1380–1398

Mergner H & Schuhmacher H (1985) Quantitative Analyse Von Korallengemeinschaften Des Sanganeb-Atolls (mittleres Rotes Meer) I Die Besiedlungsstruktur Hydrodynamisch Unterschiedlich Exponierter Außen Und Innenriffe. *Helgoländer Meeresuntersuchungen* 394: 375–417.

Miller GM et al. (2013) Increased CO_2 stimulates reproduction in a coral reef fish. *Global Change Biology* 19(10): 3037–3045

Miller KJ & Ayre DJ (2008) Protection of genetic diversity and maintenance of connectivity among reef corals within marine protected areas. *Conservation Biology* 22(5): 1245–1254

Mohamed AR & Sweet M (2019) Current knowledge of coral diseases present within the Red Sea. In *Oceanographic and Biological Aspects of the Red Sea* (pp. 387–400). Cham: Springer

Montano S et al. (2015) First record of the coral-killing sponge *Terpios hoshinota* in the Maldives and Indian Ocean. *Bulletin of Marine Science* 91(1)

Nanami A & Nishihira M (2003) Effects of habitat connectivity on the abundance and species richness of coral reef fishes: Comparison of an experimental habitat established at a rocky reef flat and at a sandy sea bottom. *Environmental Biology of Fishes* 682: 183–196

Nasr D (2015) Coral reefs of the Red Sea with special reference to the Sudanese coastal area. In *The Red Sea* (pp 453–469). Berlin and Heidelberg: Springer

Neumann AC & Moore WS (1975) Sea level events and Pleistocene coral ages in the northern Bahamas. *Quaternary Research* 5(2): 215–224

Newman WA & Dana TF (1972) *Reef Research: Regional Variation in Indian Ocean Coral Reefs:* Zoological Society of London: Symposium *No 28*. London, May 1970 D.R. Stoddart and Maurice Yonge, Eds. Published for the Society by Academic Press, New York

Nir O et al. (2014) Seasonal mesophotic coral bleaching of *Stylophora pistillata* in the Northern Red Sea. *PLoS ONE* 9(1): e84968, https://doi.org/10.1371/journal.pone.0084968

Nozawa Y et al. (2016) Seasonality and lunar periodicity in the sexual reproduction of the coral-killing sponge *Terpios hoshinota*. *Coral Reefs* 35(3): 1071–1081

Ormond RFG & Edwards AJ (1987) Red Sea fishes. *Red Sea* 7: 251

Ormond RFG et al. (1990) Test of a model of regulation of crown-of-thorns starfish by fish predators. In *Acanthaster and the Coral Reef: A Theoretical Perspective* (pp. 189–207). Berlin and Heidelberg: Springer

Osman EO (2016) *The Role of Thermal History in Shaping the Microbiome of Red Sea Corals*. Doctoral dissertation. University of Essex

Pang RK (1973) Coral reef project – *Papers in Memory of Dr Thomas F Goreau* 9 the ecology of some Jamaican excavating sponges. *Bulletin of Marine Science* 23(2): 227–243

Plucer-Rosario G (1987) The effect of substratum on the growth of *Terpios* an encrusting sponge which kills corals. *Coral Reefs* 54: 197–200.

Polunin NVC (1990) UNEP/IUCN 1988 coral reefs of the world volume 2: Indian ocean red sea and gulf UNEP regional seas directories and bibliographies IUCN Gland Switzerland and Cambridge. *Journal of Tropical Ecology* 61: 126–126

Pugh DT et al. (2019) The tides of the Red Sea. In *Oceanographic and Biological Aspects of the Red Sea* (pp 11–40). Cham: Springer

Quéré G & Nugues M (2015) Coralline algae disease reduces survival and settlement success of coral planulae in laboratory experiments. *Coral Reefs* 34(3): 863–870. http://doi.org/10.1007/s00338-015-1292-0

Rasul NM, Stewart IC, Vine P & Nawab ZA (2019) Introduction to oceanographic and biological aspects of the Red Sea. In N. Rasul & I. Stewart (Eds.), *Oceanographic and Biological Aspects of the Red Sea Springer Oceanography*. Cham: Springer. http://doi.org/101007/978-3-319-99417-8_1

Reed W (1964) The pearl shell farm at Dongonab Bay. *Sudan Notes and Records* 45: 158–163

Reimer JD et al. (2010) Domination and disappearance of the black sponge: A quarter century after the initial *Terpios* outbreak in southern. *Japan Zoology Studies* 50: 394

Reinicke GB et al. (2003) Patterns and changes of reef-coral communities at the Sanganeb-Atoll (Sudan, central Red Sea): 1980 to 1991. *Facies* 49(1): 271–297

Reopanichkul P et al. (2009) Sewage impacts coral reefs at multiple levels of ecological organization. *Marine Pollution Bulletin* 58(9): 1356–1362

Riegl BM, Berumen M & Bruckner A (2013) Coral population trajectories increased disturbance and management intervention: A sensitivity analysis. *Ecology and Evolution* 3(4): 1050–1064

Riegl BM et al. (2012) Red Sea coral reef trajectories over 2 decades suggest increasing community homogenization and decline in coral size. *PLoS ONE* 7(5): e38396

Rinkevich B (2005) What do we know about Eilat (Red Sea) reef degradation? A critical examination of the published literature. *Journal of Experimental Marine Biology and Ecology* 3272: 183–200. Web

Risk MJ & Muller HR (1983) Porewater in coral heads: Evidence for nutrient regeneration. *Limnology and Oceanography* 28(5): 1004–1008

Robitzch V & Berumen ML (2020) Recruitment of coral reef fishes along a cross-shelf gradient in the Red Sea peaks outside the hottest season. *Coral Reefs* 396: 1565–1579

Roden D (1970) The twentieth century decline of Suakin. In *Sudan Notes and Records* (vol. 51, pp. 1–22). University of Khartoum. wwwjstororg/stable/42677983

Roth F et al. (2021) High summer temperatures amplify functional differences between coral-and algae-dominated reef communities. *Ecology* 102(2): e03226

Rützler K (2004) Sponges on coral reefs: A community shaped by cooperation. *Bollettino dei Musei e degli Istituti biologici dell'Università Genova* 68: 85–148

Rützler K et al. (1993) *Terpios hoshinota* a new cyanobacteriosponge threatening Pacific reefs In: Uriz MJ & Rützler K (eds) Recent advances in ecology and systematics of sponges. *Scientia Marina* 2157: 395–403

Salam MYA (2006, July) Marine and coastal environment conservation in Sudan: The role of marine protected areas. In *Workshop on the Post-Conflict National Plan for Environmental Management in Sudan Khartoum*

Scheer G and CS Pillai (1983) *Report on the Stony Corals from the Red Sea* (V 198 pages 5 figures 41 plates 23x30cm 1100 g Language: English: Zoologica Heft 133). Schweizerbart Science Publishers Stuttgart Germany. ISBN 978-3-510-55019-7

Schönberg CHL et al. (2001) Induced colonization of corals by a clionid bioeroding sponge. *Coral Reefs* 201: 69–76.

Schuhmacher et al. (2005) The aftermath of coral bleaching on a Maldivian reef – A quantitative study. *Facies* 51(1) 80–92

Sharp KH et al (2012, August) Multi-partner Interactions in corals in the face of climate change. *The Biological Bulletin* 223: 66–77

Sheppard CA (2012) Environmental concerns for the future of Gulf coral reefs. In *Coral Reefs of the Gulf* (pp. 349–373). Dordrecht: Springer

Sheppard CA (2016) Coral reefs in the Gulf are mostly dead now but can we do anything about it? *Marine Pollution Bulletin* 1052(2016): 593–598.

Shi Q et al. (2012) Black disease (*Terpios hoshinota*): A probable cause for the rapid coral mortality at the northern reef of Yongxing Island in the South China Sea. *Ambio* 41(5): 446–455

Smith, JE, et al. (2006) Indirect effects of algae on coral: Algae-mediated, microbe-induced coral mortality. *Ecology Letters* 9: 835–845

Sneh A & Friedman GM (1980) Spur and groove patterns on the reefs of the Northern Gulfs of the Red Sea. *Journal of Sedimentary Petrology* 503: 981–986.

Spaet JLY et al. (2012) A review of elasmobranch research in the Red Sea. *Journal of Fish Biology* 805: 952–965.

Speare KE et al. (2019) Sediment associated with algal turfs inhibit the settlement of two endangered coral species. *Marine Pollution Bulletin* 144: 189–195

Srinivasan M (2003) Depth distributions of coral reef fishes: the influence of microhabitat structure, settlement, and post-settlement processes. *Oecologia* 137: 76–84

Stehli FG & Wells JW (1971) Diversity and age patterns in hermatypic corals. Systematic *Zoology* 20(2): 115–126

Syue ST et al. (2021) Testing of how and why the *Terpios hoshinota* sponge kills stony corals. *Scientific Reports* 11(1): 1–11

Tebben J et al. (2015) Chemical mediation of coral larval settlement by crustose coralline algae. *Scientific Reports* 5(1): 1–11

Ting-Ying H (1934) On the seasonal change of growth in a reef coral *Favia speciosa* (Dana) and the water-temperature of the Japanese seas during the latest geological times. *Proceedings of the Imperial Academy* 106: 353–356.

Ting-Ying H (1959) Effect of water temperature on growth rate of the reef corals/by Ting Ying H Ma. *Oceanographia Sinica: Special* 116(1): 320

Uiblein F (2021) Taxonomic review of the "posteli-species group" of goatfishes (genus *Parupeneus*, Mullidae), with description of a new species from the northern Red Sea. *Cybium* 45(1): 63–77. https://doi.org/10.26028/cybium/2021-451-008

Um N (2011) Greenlaw's Suakin: The limits of architectural representation and the continuing lives of buildings in Coastal Sudan. *African Arts* 444: 36–51.

UNEP/IUCN (1988) Coral Reefs of the World v 1: Atlantic and Eastern Pacific-v 2: Indian Ocean Red Sea and Gulf-v 3: Central and Western Pacific. UNEP/Regional Seas Directories and Bibliographies. IUCN, Gland, Switzerland and Cambridge, UK/ UNEP, Nairobi, Kenya 1 + 389pp., 36 maps

UNEP/IUCN (1988) Coral reefs of the world. Volume 2: Indian Ocean Red Sea and Gulf. UNEP/Regional Seas Directories and Bibliographies. IUCN, Gland, Switzerland and Cambridge, UK/ UNEP, Nairobi, Kenya 1 + 389pp., 36 maps

van der Ent E (2016) Abundance and genetic variation of the coral-killing cyanobacteriosponge *Terpios hoshinota* in the Spermonde archipelago SW Sulawesi Indonesia. *Journal of the Marine Biological Association of the United Kingdom* 962: 453–463.

Vermeij MJA et al. (2010) The effects of nutrient enrichment and herbivore abundance on the ability of turf algae to overgrow coral in the Caribbean. *PloS One* 5(12): e14312

Vermeij MJA et al. (2011) Crustose coralline algae can suppress macroalgal growth and recruitment on Hawaiian coral reefs. *Marine Ecology Progress Series* 422: 1–7

Veron J (2013) Overview of the taxonomy of zooxanthellate Scleractinia. The Linnean Society of London, *Zoological Journal of the Linnean Society* doi: 10.1111/zoj.12076

Vine PJ (1970) Field and laboratory observations of the crown-of-thorns starfish *Acanthaster planci*: Densities of *Acanthaster planci* in the Pacific Ocean. *Nature* 228: 341–342. https://doiorg/101038/228341a0

Vine PJ (1972a) Spirorbinae (Polychaeta: Serpulidae) from the Red Sea, including descriptions of a new genus and four new species. *Zoological Journal of the Linnean Society* 51(2): 177–201

Vine PJ (1972b) Coral-reef conservation around the Seychelles Indian ocean. *Biological Conservation* 4(4): 304–305

Vine PJ (1973) Crown of thorns (*Acanthaster planci*) plagues: The natural causes theory. *Atoll Research Bulletin* 166: 1–10. https://doiorg/105479/si007756301661

Vine PJ (1974) Effects of algal grazing and aggressive behaviour of the fishes *Pomacentrus lividus* and *Acanthurus sohal* on coral-reef ecology. *Marine Biology* 242: 131–136.

Vine PJ (2019) Red Sea research: A personal perspective. In *Oceanographic and Biological Aspects of the Red Sea* (pp. 215–237). Cham: Springer

Vine PJ & Head MS (1977) Growth of corals on Commander Cousteau's underwater garage at Shaab Rumi (Sudanese Red Sea). *Journal of the Saudi Arabian Natural History Society* 1977: 6–18

Vine PJ & Vine MP (1980) Ecology of Sudanese coral reefs with particular reference to reef morphology and distribution of fishes. *Proceedings of a Symposium on the Coastal and Marine Environment of the Red Sea Gulf of Aden and Tropical Western Indian Ocean* (Vol. 1) UNESCO/Sudanese Scientific Research Group/University of Khartoum

Wear SL & Thurber RV (2015) Sewage pollution: Mitigation is key for coral reef stewardship. *Annals of the New York Academy of Sciences* 1355

Webster NS et al. (2013) Ocean acidification reduces induction of coral settlement by crustose coralline algae. *Global Change Biology* 19(1): 303–315

Weis VM (2008) Cellular mechanisms of Cnidarian bleaching: Stress causes the collapse of symbiosis. *The Journal of Experimental Biology* 211(19): 3059–3066. https://doi.org/10.1242/jeb.009597

Wild C et al. (2014) Turf algae-mediated coral damage in coastal reefs of Belize, Central America. *PeerJ* 2: e571

Wilmes JC (2018) Contributions of Pre- versus Post-settlement processes to fluctuating abundance of crown-of-thorns starfishes (*Acanthaster* spp). *Marine Pollution Bulletin* 135: 332–345.

Wilson SK et al. (2006) Multiple disturbances and the global degradation of coral reefs: Are reef fishes at risk or resilient? *Global Change Biology* 12(11): 2220–2234

Yao F et al. (2014) Seasonal overturning circulation in the Red Sea: 1 Model validation and summer circulation. *Journal of Geophysical Research: Oceans* 119(4): 2238–2262

Yara Y et al. (2011) Projection and uncertainty of the poleward range expansion of coral habitats in response to sea surface temperature warming: A multiple climate model study *Galaxea. Journal of Coral Reef Studies* 13(1): 11–20

Yuasa H et al. (2017) Diet of *Acanthaster brevispinus* sibling species of the coral-eating crown-of-thorns starfish *Acanthaster planci* sensu lato. *Bulletin of Marine Science* 93(4): 1009–1010

Index

Note: Page numbers in *italics* denote figures or tables on the corresponding page.